Seismic Data Processing with Seismic Un*x

A 2D Seismic Data Processing Primer

**David Forel, Thomas Benz,
and Wayne D. Pennington**

Michigan Technological University

Course Notes Series No. 12
Lawrence Gochioco, Series Editor

Society of Exploration Geophysicists
Tulsa, Oklahoma, USA

ISBN 0-931830-48-6 (Series)
ISBN 1-56080-134-4 (Volume)

Society of Exploration Geophysicists
P.O. Box 702740, Tulsa, OK 74170-2740, USA

Published in 2005
Reprinted in 2006, 2007, 2008, 2010

Printed in the United States of America

Short Table of Contents

Table of Contents

Preface

This book can serve either of two purposes. (1) It can be used, as it is in our courses at Michigan Technological University, as an aid to teaching seismic reflection data processing. (2) It can be used as a primer to Seismic Un*x (SU) by those who may or may not already be familiar with seismic processing using other software packages. SU is provided by the Center for Wave Phenomena at Colorado School of Mines and is available from their web site www.cwp.mines.edu/cwpcodes.

There are details of SU that are important to the processing specialist, but are not essential to the student who is being introduced to seismic processing. Where these details appear in the text, we placed a gray bar in the margin to indicate that the material can be skipped by the student interested only in learning about processing. Of course, these details might be among the most-interesting material from the viewpoint of someone who wants to learn programming techniques that apply to SU.

A beginning course in reflection seismology processing might complete the entire book in one semester. A course that covers fundamental processing and some interpretation might skip the material with a gray bar and use only one of the real (field, not synthetic) data sets that we provide.

We have found that one of the biggest hurdles to developing a new course is the scarcity of real data. With this book, we provide two real data sets. The first real data set, from the Nankai trough near Japan, is unusual in that it was acquired over very deep water. This presents some advantages (mediocre velocity analyses will still produce good results), but it is not realistic for typical exploration purposes. It is, however, a site of exciting geologic features, and the rugged seafloor topography dramatically demonstrates the benefit of migration. It is a data set that any student can appreciate. The second real data set, from offshore Taiwan, presents a number of processing challenges; it is much more difficult to process to satisfaction.

The compact disks (CDs) included with this book have copies of the scripts and the seismic data sets (including some data sets that are generated by the scripts). The book is printed in black-and-white to keep its price low. Because some figures are best viewed in color, we put copies of all color figures, in uncompressed TIFF format, on the CDs. Appendix E has a complete list of the contents of the CDs.

We at Michigan Tech will maintain a list of errata for the book. The errata list can be found by searching on the web for Michigan Tech, errata, and Seismic Un*x. Please report any errors to the address identified on that web site. Reports of SU bugs, suggestions for SU improvements, or proposals for new SU scripts should be submitted to the CSM Center for Wave Phenomena via methods suggested on their web site.

Table of Computer Notes

Computer Notes are guides to help you better understand and use the Unix system. Chapter 2 is not listed here, but it can be considered a chapter of Computer Notes.

Warning 1: We developed this Seismic Un*x Primer while using the *csh* shell. The Computer Notes were written from this perspective. If you are not working under the *csh* shell, our Computer Notes might not be appropriate for you.

Warning 2: The previous warning does not apply to the scripts. The first line of every script is, "`#! /bin/sh`", a command that makes scripts run under the Bourne shell.

Throughout this Primer, we use a vertical bar, as shown to the right, to mark passages that a first-time reader can safely ignore. Generally, the bar marks two kinds of text. Details of model building (large portions of Chapters 4, 5, and 6) can be ignored because Model 4 is on one of the CDs. The bar also marks parts of scripts and explanations of those parts that are of more interest to shell script programmers than to script users.

Table of Scripts

Scripts are listed when introduced. The page number is the page of the first line of the script. When a script "fragment" is listed, the original script location is also cited.

Line commands are not referenced, with the exception of our only use of **sumute**.

1. Introduction

1.1 The Goal of this Primer

Our objective is to introduce you to the fundamentals of seismic data processing with a learn-by-doing approach. We do this with Seismic Un*x (SU), a free software package maintained and distributed by the Center for Wave Phenomena (CWP) at the Colorado School of Mines (CSM). At the outset, we want to express our gratitude to John Stockwell of the CWP for his expert counsel.

SU runs on several operating systems, including Unix, Microsoft Windows, and Apple Macintosh. However, we discuss SU only on Unix.

Detailed discussion of wave propagation, convolution, cross- and auto-correlation, Fourier transforms, semblance, and migration are too advanced for this Primer. Instead, we suggest you refer to other publications of the Society of Exploration Geophysicists, such as "Digital Processing of Geophysical Data - A Review" by Roy O. Lindseth and one of the two books by Ozdogan Yilmaz: "Seismic Data Processing," 1987 and "Seismic Data Analysis," 2001.

Our goal is to give you the experience and tools to continue exploring the concepts of seismic data processing on your own.

1.2 The Outline of this Primer

This Primer covers all processing steps necessary to produce a time migrated section from a 2-D seismic line. We use three sources of input data:

- Synthetic data generated by SU;
- Real shot gathers from the Oz Yilmaz collection at the Colorado School of Mines (ftp://ftp.cwp.mines.edu/pub/data); and
- Real 2-D marine lines provided courtesy of Prof. Greg Moore of the University of Hawaii: the "Nankai" data set and the "Taiwan" data set.
    - o The University of Texas, the University of Tulsa, and the University of Tokyo collected the Nankai data. The U.S. National Science Foundation and the government of Japan funded acquisition of the Nankai data.
    - o The University of Hawaii, San Jose State University, and National Taiwan University collected the Taiwan data. The U.S. National Science Foundation and the National Science Council of Taiwan funded acquisition of the Taiwan data.

Chapters 1-3 introduce the Unix system and Seismic Un*x.

Chapters 4-5 build three simple models (complexity slowly increases) and acquire a 2-D line over each model. (These chapters may be skipped if you are only interested in processing.)

Chapters 6-9 build a model based on the previous three, acquire a 2-D line over that model, and process the line through migration.

Chapters 10-11 start with a real 2-D seismic line of shot gathers (Nankai) and process it through migration.

Chapters 12-13 and 15-16 start with a real 2-D line of shot gathers (Taiwan) and process it through migration.

Chapter 14 uses real shot gathers from the Oz Yilmaz collection to demonstrate f-k filtering and deconvolution.

We introduce the major processing steps in the following chapters:

Build a model	Chapter 4
Acquire seismic data	Chapter 5
Amplitude correction	Chapter 10
Frequency (1-D) filtering	Chapter 10
Frequency-wavenumber (f-k) filtering	Chapter 14
Deconvolution	Chapter 14
Sort from shot gathers to CMP gathers	Chapter 7
Velocity analysis	Chapter 7
Normal moveout correction (NMO)	Chapter 8
Stack	Chapter 8
Migration	Chapter 9

Scripts are included for each process. A detailed description of these scripts is the heart of this Primer.

Experience with the Unix operating system is helpful, but not necessary. Chapter 2 introduces Unix commands and SU scripts.

1.3 A Bit about Seismic Un*x

Seismic Un*x (Cohen & Stockwell, 2002) is a free software package for seismic data processing. SU is maintained and regularly updated by the Center for Wave Phenomena (CWP) at the Colorado School of Mines (CSM). The SU home page at the CWP is:

http://www.cwp.mines.edu/cwpcodes/

You can download SU from the CWP ftp server and install it on almost any Unix system. The source code is also included, making it possible for you to adjust and expand SU. The CWP web site also contains links to other useful web sites.

Please read the legal statement that accompanies SU. For your convenience, we make this available as our Appendix A.

SU is easy to use because it does not require extensive programming knowledge, only basic Unix commands for file management, pipes, and shell redirection (see Chapter 2).

Although SU can be run from the command line, the most efficient way to use SU is through shell scripts. Shell scripts let us combine several processing programs into one "job," similar to commercial seismic processing packages. We first describe Shell scripts in Chapter 2.

1.4 Computer Note 1: Enter

When we tell you to "enter" something on the computer, we want you to type the instruction, then press the **Enter** or the **Return** key. Pressing **Enter** or **Return** tells the computer to execute the instruction. Merely typing an instruction on a line does not execute the instruction.

1.5   Computer Note 2: Your Shell

When you enter a command, you are interacting with the Unix shell. There are several Unix shells: the Bourne shell, the C shell, the Korn shell, the Bourne Again shell (BASH), and others. **All example scripts in this Primer use the Bourne shell; most of the Computer Notes refer to the C shell.**

1.6   Computer Note 3: Changing Directories

1. Open an x-term window and enter **pwd** (on some computers you have to use **cwd**) to see the directory in which you are now working.

2. Enter **ls** to see a listing of the files and subdirectories in that directory.

3. If you have a subdirectory; for example, "work," enter **cd work** to change to that subdirectory, then enter **ls** to see a listing within it.

4. When you want to move up one level of directory, enter **cd ..** (Yes, that is: cd space dot dot.)

5. If you enter **cd** by itself, you will go to your home directory.

6. If you enter **cd .** (cd space dot), you will not go anywhere because the single dot (period) is a reference to your current directory.

    The single dot can be useful. Suppose there is a system calculator called xcalc and, suppose you wrote your own calculator program that is also called xcalc. If you enter

        xcalc

    the system version might start instead of yours. To run your version instead of the system version, go to the directory that contains your version, then enter

        ./xcalc

    The dot slash tells the system to look in the current directory for the program.

1.7   Computer Note 4: Cursor Prompt

Usually, the default computer prompt does not give much information. For example, Figure 1.1 shows the default prompt when I (Forel) open a new x-term window

Figure 1.1: *Forel's default x-term prompt.*

The prompt of Figure 1.1 shows the name of the machine that I am on and the number of the command (1) that I am about to enter. I think a more useful prompt is one that tells me which directory I am in (Figure 1.2).

A disadvantage of a prompt that shows the current directory is that as I go to deeper subdirectories, the prompt becomes longer (Figure 1.3). I am willing to pay this price for the information that I gain.

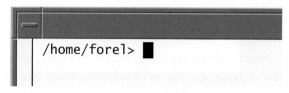

Figure 1.2: *My new prompt shows the current directory I also changed "%' to ">".*

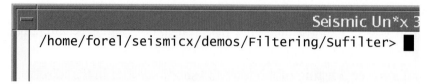

Figure 1.3: *A prompt with a long path name.*

To change your C shell prompt, edit your home directory's shell resource file. The file ".cshrc" is the resource file for the C shell. Yes, the file name starts with a period (.). Use any editor you like to add the following lines to your ".cshrc" file.

Note: We use only Unix-based text editors to write and edit shell scripts and resource files. Many personal computer editors insert hidden format codes that might be wrongly interpreted by the Unix operating system.

```
# Make prompt contain "pwd"
alias setprompt 'set prompt="$cwd> "'
alias cd 'cd \!* && setprompt'
alias pushd 'pushd \!* && setprompt'
alias popd 'popd \!* && setprompt'
setprompt
```

The first line is a comment. When the first character is "#," the entire line is ignored.

After you save your changes, when you open a new x-term window, the system will read the resource file and use those instructions in the new x-term window. If you want to use .cshrc changes in your current window, enter:

```
source   .cshrc
```

1.8   Your First Seismic Un*x Instruction

Before you can run Seismic Un*x, you must be sure your SU commands will be interpreted by the Unix operating system. This means SU must be "in your path." Your system administrator may have made SU part of your login path or you may have to enter a command first. The approach we use at Michigan Tech is described in Appendix B.

Let's run a simple program to test that you and your SU environment are working together. Enter (type the following, then press the **Enter** or the **Return** key):

```
suplane  |  suxwigb  &
```

You should get a new window with a picture of some synthetic seismic data on your screen (Figure 1.4). If you do not, ask someone for help.

The style of the above command is:

```
[program]  ["pipe" output data to next program]  [program]  &
```

The pipe "|" directs data from one process to another.

You can regain the Unix prompt in your SeismicX window by pressing the return key (if you supplied **&** at the end of the command). You can close the graphics window by using the Unix **Close** command (the top left button).

Let's discuss the details of the above command:

- **suplane** creates seismic data.
- the pipe (the vertical bar) directs (sends) the data to the next program.
- **suxwigb** (the X-window display program) displays the seismic data on your screen as wiggle traces.
- The **&** needs a special explanation. Read the next section.

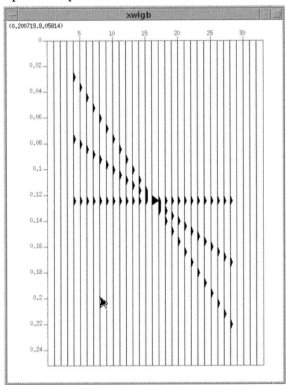

Figure 1.4: *The output of your first SU command. The upper-left corner shows (0.200719,8.05814) because we clicked the middle mouse button at approximately 0.200 ms time (the vertical axis is time) and on the eighth trace from the left (the horizontal axis is trace number).*

1.9 Computer Note 5: Background (&) and Foreground Processes

1.9.1 Background Processes

In technical language, using **&** (ampersand) at the end of a command runs (spawns) the process in the background. In simple terms, using **&** does not lock your cursor. Since your cursor is not locked (frozen) when you use **&** at the end of a command, you can use the same x-term to do other things, like submit other processes. For example, the following command starts the Netscape web browser as a background process:

```
netscape &
```

To enter other commands after this one, press Enter again to regain a prompt, then enter your next command.

While a process is running in the background, you might need to cancel (kill) it. You can do this using a Unix utility called **top**.

1. In an x-term window, enter **top**
2. The x-term window now shows your most active processes (Figure 1.5) and the cursor is now in the upper left of the **top** display (shown as the red box in Figure 1.5).

```
load averages:  0.01,  0.03,  0.05
73 processes:  72 sleeping, 1 on cpu
CPU states: 97.2% idle,  0.4% user,  2.4% kernel,  0.0% iowait,  0.0% swap
Memory: 384M real, 107M free, 131M swap in use, 2157M swap free

  PID USERNAME THR PRI NICE  SIZE   RES STATE  TIME    CPU COMMAND
 8731 forel     11  48    0   18M   16M sleep  1:16  0.37% dtmail
 9811 forel      4  59    0   19M   14M sleep  0:01  0.19% sdtimage
 6942 root       1  59    0  144M  101M sleep  5:26  0.15% Xsun
 9809 forel      1  49    0 2240K 1280K cpu    0:00  0.14% top
 9828 forel      1  25    5   26M   18M sleep  0:04  0.00% netscape
```

Figure 1.5: *Screen capture of Forel's* **top** *processes.*

3. The left column (PID) identifies process netscape, the Netscape web browser. as 9828. To kill this process, enter

   k 9828

4. To exit **top**, enter **q** for quit.

Note: You can use **top** to learn the process ID (PID), then kill the process using a line command. To kill the Netscape process described above using a line command, enter:

   kill -9 9828

The flag "-9" tells **kill** that you are serious about killing process 9828.

1.9.2 Foreground Processes

When you start a process and don't use **&**, the new process uses the current x-term (the process runs in the foreground), which locks the cursor. For example, the following command starts the Netscape web browser as a foreground process:

   netscape

To continue working when the cursor is locked, you must either kill the process that you just started or you must open a new x-term. To kill a foreground process in an x-term window:

1. click the window to make it active,
2. hold down the **Control** key and briefly press the **c** key (press **Control-c**).

This kills the process. Because the process dies, the browser window also closes.

1.10 Zoom the xwigb Window

You can zoom the image in the **xwigb** window. Put the cursor somewhere in the **xwigb** window, press and hold down the left mouse button, drag the mouse until the box encloses the area you want enlarged, then let go of the button.

You can zoom repeatedly. Notice that time numbers and trace numbers correspond to the zoomed data. Click once in the window to return to the original view.

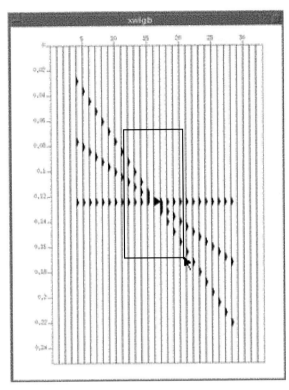 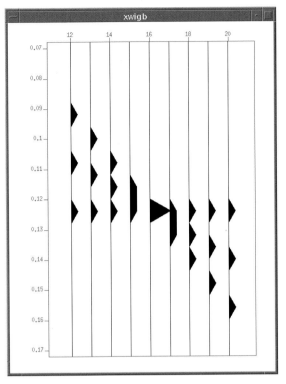

Figure 1.6: *Zoom an **xwigb** window by dragging a box with the left mouse button. The left image shows the mouse dragging the box. The right image shows the zoomed image. To return to the original view, click the left mouse button in the window.*

1.11 The Demos Directory

The **demos** directory is a standard part of the SU installation. If you don't know where yours is, ask someone. It is full of easy to use demonstrations of many seismic data processing programs.

Once you are in the **demos** directory, repeatedly use **cd** to move around. Use **ls** and **pwd** frequently to see where you are. Notice the many README files. The README files tell how to use the demonstrations in each directory. You can write README files to your screen by typing either of the following commands:

```
cat   README
more  README
```

If you use the second command, you can advance the text that appears on your screen by pressing the spacebar to move to the next page. Pressing **Enter** or **Return** scrolls the display one line.

Exploring (executing) the demonstrations is an easy way to see seismic data processing!

1.12 Self Documentation in Seismic Un*x

To display documentation for programs like **suplane** or **suxwigb**, enter the name of the program without arguments. "Without arguments" means you enter only the name of an SU program. For example, enter

```
suplane
```

SU interprets this to mean you want information about **suplane**, so SU prints the program help (selfdoc) file to the screen. Below are the first several lines of the **suplane** selfdoc file.

```
SUPLANE - create common offset data file with up to 3 planes

suplane [optional parameters] >stdout

Optional Parameters:
 npl=3                          number of planes
 nt=64                          number of time samples
 ntr=32                         number of traces
 taper=0                        no end-of-plane taper
                                = 1 taper planes to zero at the end
 offset=400                     offset
 dt=0.004                       time sample interval in seconds
```

You see that the name stands for "create common offset data file with up to 3 planes." (Many SU program names start with "su.") You also see the default values (values used when the user does not supply a value) of various parameters, such as the number of planes (*npl=3*), the number of time samples per trace (*nt=64*) and the time sample interval (*dt=4 ms*). Using default values, each trace is 256 ms long.

1.13 Your Second Seismic Un*x Instruction

Let's change the command given earlier in Section 1.8. Enter:

suplane   npl=2   dt=0.008 | suxwigb   &

By supplying values for *npl* and *dt*, you overwrite (replace) **suplane**'s default values. Your output (pictured below) has only two planes and each trace is 512 ms long.

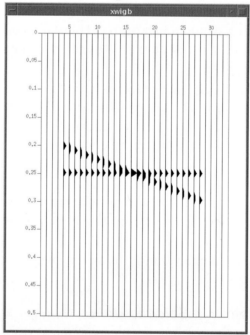

Figure 1.7: *The **suplane** command with two defaults overwritten: number of planes = 2, and time interval = 0.008 seconds.*

1.13.1 Computer Note 6: No Spaces around the Equal Sign

Do not use spaces before or after the equal (=) sign

1.14 *Other Help Facilities in Seismic Un*x*

Various help facilities are described in The New SU User's Manual by John W. Stockwell, Jr. & Jack K. Cohen; Version 3.2: August 2002. This Manual is available at the CWP Seismic Un*x web site. In addition to the self documentation described in Section 1.11, other help facilities include:

- **suhelp** lists all the available programs.
- **suname** lists all programs and libraries with a short description of them.
- **sudoc** followed by the program name gives documentation of the program. This can work even when there is no selfdoc for the program.
- **sufind** followed by a string searches in all self documentation for this string.
- **sukeyword** lists keys used for headers.
- **demos** is a directory in the SU installation that contains useful scripts.

2. Unix Commands and Concepts

2.1 Elementary Unix Commands

The following table lists essential Unix commands.

Table 2.1: Unix commands

Command	Description
|	The "pipe" connects two processes.
&	Run (spawn) a process in the background.
cat	Display a text file on the screen. Example: **cat zebra.txt**
cd	Change to home directory **or** change to directory specified after space.
cd ..	Go up one directory
chmod	Change file permissions mode. Example: **chmod +x filter.su**
cp	Make a copy of a file. Example: **cp zebra2.txt zebra3.txt**
find	Search for (find) files within and below a directory. Example: **find . -name 'acq*.sh' -print**
ls	List directories and files.
ls -a	List directories and files, including hidden files (files with a dot prefix).
ls -lF	List directories and files using long format.
man	Access Unix (not Seismic Un*x) help (manual) pages. Example: **man ls**
mkdir	Make directory. Example: **mkdir data**
more	Display the contents of a text file, one screen at a time. If the file occupies more than one screen, press the space bar to scroll a page or the Return (Enter) key to scroll a line.
mv	Move or rename a file or directory. Example: **mv zebra2.txt zebra3.txt**
pwd (or cwd)	Show "present working directory."
rm	Remove (delete) a file. Example: **rm zebra.txt**
rmdir	Remove a directory. (A directory must usually be empty before it can be removed.) Example: **rmdir data**
Control-c	Kill (stop) a process using the active window.

This chapter explains fundamental Unix commands that are necessary for understanding later scripts. It will be helpful if you know elementary Unix commands. Books titled "Teach Yourself Unix" have excellent, simple, early chapters that give the basics. Also, by surfing the web, you can find universities that have good tutorial sites.

2.2 File Name Conventions

The following file name suffixes are used throughout this Primer.

- .sh   shell script
- .scr   shell script that launches a .sh shell script
- .su   binary seismic data
- .dat   binary data, not seismic data
- .eps   image file formatted as Encapsulated Postscript (EPS)
- .txt   ASCII data file

The SU seismic data format is based on, but is not identical to, the binary format called SEG-Y. SEG-Y was defined by the Society of Exploration Geophysicists (SEG) and has become an industry standard format for seismic data exchange.

The following file types can be printed directly to the screen by the **cat** and **more** commands: .sh, .scr, .txt.

2.3 Advanced Unix Concepts

We do not explain the following advanced Unix concepts in detail. Their usage will be made clear by the way we use them in later scripts.

The following Unix commands are not complete. These are merely a selection of commands that we consider helpful for your understanding and reproduction of the processing in this Primer.

Remember that Unix is case sensitive. That is, **suplane** is not the same as Suplane.

2.3.1 Invoke the Shell: `#! /bin/sh`

To start the Bourne shell interpreter, the first line of any script we make must be:

```
#! /bin/sh
```

2.3.2 Comment Line:

A comment line is one that is not seen by the shell. It is for user-comments within a shell script. A comment line begins with the # character. For example:

```
# Name input and output files
```

Note: The use of "#" shown in the previous section is a rare exception to the use of "#" explained here.

Note: The shell ignores blank lines.

Note: Do **not** insert comment lines or blank lines in the middle of an instruction.

2.3.3 Pipe: |

We saw the pipe in Section 1.8. We mention it here because it is an advanced concept. A pipe allows data to flow from one process to another. Below (as before),

suplane creates data, then the pipe sends the data to the imaging program, **suxwigb** to be seen on the screen. In other words, data flow is through the pipe, left to right.

```
suplane  |  suxwigb
```

2.3.4 Shell Redirection: `<` and `>` and `>>`

The redirect **<** sends data from right to left. Below, data file *seis1.su* is sent as input to program **suwind**, a windowing program that we will use later.

```
suwind  <  seis1.su
```

The redirect combination below sends data file (a) to process (b); the output of (b) is stored in file (c).

```
suwind  <  seis1.su  >  seis2.su
  b     <     a      >    c
```

The first command below types the contents of file *README* to the screen. The second command below redirects the result of the **cat** command to file *info2.txt*; the contents of file *README* do not get to the screen. The third command below appends the contents of file *README* to the bottom of file *info2.txt*; that is, the original contents of file *info2.txt* are not overwritten.

```
cat README
cat README  >  info2.txt
cat README  >>  info2.txt
```

2.3.5 Permissions: `rwxrwxrwx`

All files on a Unix system have permission rights associated with them. These rights of "read" (r), "write" (w), and "execute" (x) are assigned to "user," "group," and "other." You can view the permissions of a file by entering **ls -l**. (We prefer **ls -lF**.) The file permissions are in the left column. For example:

```
/home/forel/suscripts> ls -lF
total 160
-rw-r--r--   1 forel    geograd      1915 Apr 15 19:02 nmoall.sh
drwxr-xr-x   2 forel    geograd       512 Apr 15 11:08 data/
```

Permissions can be changed with the command **chmod** (change the permissions mode of a file). ***SCRIPT FILES MUST BE EXECUTABLE.*** To make a file executable, enter **chmod +x** and the file name. For example:

```
/home/forel/suscripts> chmod +x nmoall.sh
/home/forel/suscripts> ls -lF
total 160
-rwxr-xr-x   1 forel    geograd      1915 Apr 15 19:02 nmoall.sh*
drwxr-xr-x   2 forel    geograd       512 Apr 15 11:08 data/
```

The left column shows that file *nmoall.sh* now is executable. The "*" on the file name (right column) also indicates that the file is executable.

2.3.6 Shell Internal Variables

Inside a shell script, variables are defined with an equal (=) sign and referenced (used in the script) with the $. The line

```
offset=2000
```

sets the variable *offset* to 2000. The value of the variable is later used as *$offset*.

2.3.7 Passing Values to the Shell

Values can be passed to the shell when the script is called. Each argument listed after the name of the script when it is called can be referenced in the script with the expressions $1 (first argument), $2 (second argument), etc. The command

```
migrate1.sh  1800  4500
```

starts script *migrate1.sh* and assigns the value 1800 to variable *$1* and the value 4500 to variable *$2*. If *migrate1.sh* is a migration script, 1800 might mean a velocity of 1800 m/s, the velocity at the top of the time section; and 4500 might mean 4500 m/s, the velocity at the bottom of the time section. By passing values to the shell when you start the script, you can re-run the script any number of times with different values, without rewriting the script, until you are satisfied with the output.

2.3.8 Continuation Mark: \

Sometimes we place instructions on more than one line for the sake of visual clarity. In the following instruction, the title of the output plot is placed on the second line. When the continuation mark (backslash) is used, no other character or space can be to its right.

```
suximage < shot$1.su perc=95 \
          title="SP # $1"
```

2.3.9 Integer Evaluation

Integers are numbers without a decimal part. That is, 2, 0, and -1 are integers. Simple integer expressions can be evaluated within the shell by enclosing the expression in single back quotes (***not the apostrophe***). Notice that the multiply sign must use the backslash.

```
var=`expr var1 [+][-][\*][/][%] var2`
```

For example, increment a loop counter with:

```
k=`expr $k + 1`
```

There are no spaces around the equal sign (=). However, spaces are acceptable within the quotes.

2.3.10 Floating Point Evaluation

Floating point numbers (sometimes called "floats") are numbers with a decimal part. That is, 2., 0.01, and -1.5 are floating point numbers. Floating-point arithmetic is possible using the Unix calculator **bc**. For example:

```
src=`bc -l << -END
$i * 0.05
END`
```

Here, *src* is the variable. The value of *src* is computed in the second line. Most of the first line and the third line are **bc** syntax that must be used exactly as shown.

Note: Strictly speaking, the characters on the third line must be the only characters on the line and must start in column 1. However, modern operating systems are not as strict as older ones. Scripts in this Primer ignore this outdated, strict rule. However, beware!

More sophisticated script languages like *awk* or Perl could also be used to create scripts with floating point arithmetic, but this would make the use of SU unnecessarily complicated in the context of this Primer.

2.3.11 Debug Option

Used as the first command of a script after initializing the shell, the debugging command prints all executed lines to the screen as they execute. The screen output can be obscure, but it can also be helpful. The command is:

```
set  -x
```

2.3.12 Case Option

Case involves pattern matching. If *pattern1* matches input variable *var*, commands in *list1* are executed. If, instead, *pattern2* matches *var*, commands in *list2* are executed. Etc.

```
case [var] in
    [pattern1]) [list1];;
    [pattern2]) [list2];;
        ...         ...  ;;
    esac
```

An important part of *case* is the right parenthesis after the key words pattern1, pattern2, etc. Also, notice the double semicolons at the end of each *list*.

2.3.13 Shell Loops and If Blocks

The programming language of the Bourne shell allows making *for* loops:

```
for [var]
do
    [list]
done
```

do-while loops:

```
while [condition]
do
    [list]
done
```

and *if*-blocks:

```
if [condition]
then
    [list1]
else
    [list2]
fi
```

2.3.14 Exit the Shell

At the end of a script the command **exit** can be used. If **exit** is used, the shell terminates at this point, even if there are commands after **exit**. The command is:

```
exit
```

2.4   Shell Script Example: myplot.sh

This example shows the syntax of a "case" script and demonstrates SU plot options. We placed line numbers on the left for discussion; they are not part of the script.

```
1   #! /bin/sh
2   # File: myplot.sh
3   #       Plot test of "case"
4
```

```
 5  # Set messages on
 6  ##set -x
 7
 8  # Define variable
 9  signaltonoise=10
10
11  # Create seismic data. By default, suplane generates
12  # 32 traces with signals from three reflectors.
13  suplane | suaddnoise sn=$signaltonoise > myplot.su
14
15  # Send file myplot.su to user-selected image program
16  # or make Postscript file.
17  case $1 in
18
19    wiggle)
20      suxwigb   < myplot.su title="Wiggle plot"
21      ;;
22
23    image)
24      suximage  < myplot.su title="Bitmap plot"
25      ;;
26
27    pswiggle)
28      supswigp  < myplot.su > myplot1.eps title="Postscript Wiggle"
29      echo " "
30      echo "  Wiggle file  myplot1.eps  has been created."
31      echo " "
32      ;;
33
34    psimage)
35      supsimage < myplot.su > myplot2.eps title="Postscript Bitmap"
36      echo " "
37      echo "  Bitmap file  myplot2.eps  has been created."
38      echo " "
39      ;;
40
41    *)
42      echo " "
43      echo "  Use: myplot.sh [wiggle, image, pswiggle, psimage]"
44      echo " "
45      ;;
46
47  esac
48
49  # Exit politely from shell
50  exit
51
```

(For curious readers, we placed numbers in the left by entering:

```
cat  -n  myplot.sh  >  myplot.txt
```

Option –*n* of **cat** made the numbers. We redirected the **cat** output to a text file that we later inserted into this document.)

Line 1 starts the shell. Line 2 is the first comment. Line 9 defines the signal-to-noise ratio of the output seismic data. The defined variable *signaltonoise* is referenced in line 13 as $*signaltonoise*. (Refer to Section 2.3.6.)

Line 13: program **suplane** creates seismic data, the pipe (|) sends the output of **suplane** to program **suaddnoise**, program **suaddnoise** adds the previously defined amount of white (random) noise, the modified seismic data is redirected (>) to output as file *myplot.su*. By the end of line 13, you have a new file of seismic data on your computer. File *myplot.su* is the input data to the "case" logic that follows.

Lines 17-47 are the "case" logic.

Line 50 exits the shell.

To run this script, it first must be created in an editor, saved under any name and made executable. The script is then called with the name of the file, followed by the case selection (see Table 2.2). For example, if the script file was saved under the name *myplot.sh*, and then made executable with the command

```
chmod  +x  myplot.sh
```

the command

```
myplot.sh  wiggle
```

runs the script and displays wiggle traces on the screen. In fact, there are four ways to run this script:

Table 2.2: Possible cases and output of *myplot.sh*

Command	Output
`myplot.sh  wiggle`	wiggle trace screen image
`myplot.sh  image`	bitmap screen image
`myplot.sh  pswiggle`	wiggle trace Postscript file – *myplot1.eps*
`myplot.sh  psimage`	bitmap Postscript file – *myplot2.eps*

The first two cases display seismic data directly on the screen. The latter two cases save the created image as a Postscript file. When you run the script selecting one of the Postscript cases, you must use one of the commands below to see the contents of the Postscript files (if you have the Ghostscript program):

```
ghostview  -bg white  myplot1.eps &
ghostview  -bg white  myplot2.eps &
```

Your output should look like one of the four images in Figures 2.1 and 2.2 (below). Option "-bg white" makes the ghostview background white. On many systems, the default background is grey.

The **echo** command writes to the screen (unless the output of **echo** is redirected). Lines 29 and 31 put blank lines above and below the output of line 30 to make line 30 output easier to read. The same is true for lines 36, 38, 37 and lines 42, 44, 43.

Last, notice line 41. This case option is used if you do not select or if you incorrectly type one of the other cases. Case option "*)" is placed after all other cases and is always selected if none of the previous cases are selected. This option is used here to print information to the screen to remind you of the acceptable cases.

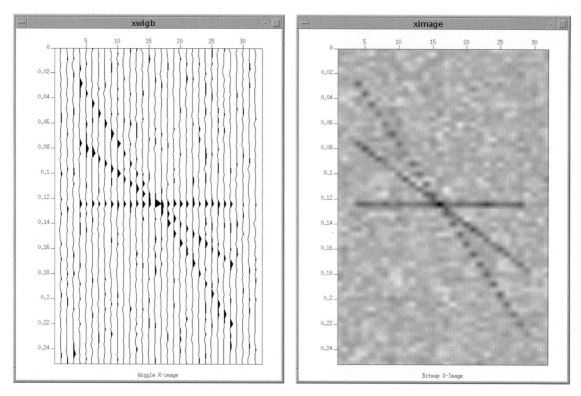

Figure 2.1: *Left: case = wiggle. Right: case = image.*

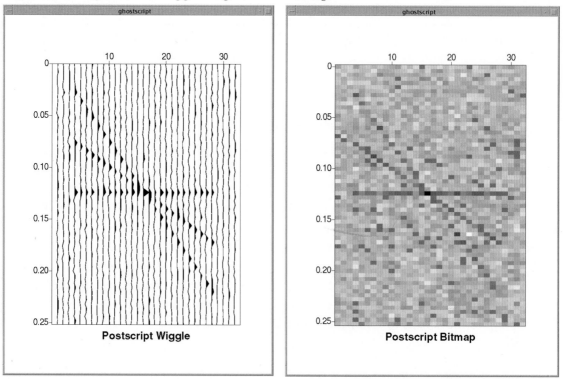

Figure 2.2: *Left: case = pswiggle. Right: case = psimage.*

3.   Trace Headers and Windowing Data

3.1   What are Trace Headers?

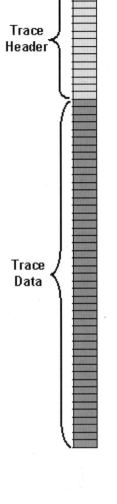

Trace Header

Trace Data

When seismic traces are in SEG-Y format, the SU trace format, and many other formats, the beginning of every trace, the trace header, has information about the trace. You can think of these as slots of information above the data part of the trace. The data part is the time-amplitude series that we see in a seismic display.

Trace header information might include the trace number and the offset of the trace (for shot or CMP gathers). In an SU seismic data set, the number of trace header slots is the same to ensure that every trace in a data set has the same length (in terms of bytes of storage).

SU doesn't call them headers; it calls them keys. The following table lists some SU keys.

Table 3.1: Some SU trace "headers" or trace "keys"

Key	Definition
dt	sample interval in microseconds
ns	number of samples in this trace
ntr	number of traces
offset	offset
tracf	trace number within field record (gather)
tracl	trace sequence number within line
tracr	trace sequence number within reel (entire data set)
delrt	delay recording time in milliseconds

3.2   First Look at Trace Headers (keys): surange and sugethw

You can use program **surange** to learn which keys are in a data set and the range of their values (the largest and smallest values). Now that you have run script *myplot.sh* (Section 2.4), use program **surange** to examine the trace headers of the .seismic file. Enter:

```
surange  <  myplot.su
```

The output (screen display) is:

```
32 traces:
 tracl=(1,32)  tracr=(1,32)  offset=400 ns=64 dt=4000
```

This synthetic data set has 32 traces. Only five of its keys (trace headers) have non-zero values. The minimum value of *tracl* and *tracr* is 1; the maximum value of *tracl* and *tracr* is 32. This is not surprising since there are 32 traces in the data set. We can suppose that the traces are numbered in sequence, 1 to 32. Surprisingly, all traces have the same offset: 400.

The sample interval is 4 ms (*dt = 4000* microseconds) and every trace has 64 samples (*ns = 64*). So, the traces must be 256 ms long. We can see this is true in Figure 1.4.

Let's use another trace header analysis program, **sugethw**, a program that gets header words and writes them to the screen. The command below sends seismic data file *myplot.su* to program **sugethw**. Keys *tracl*, *tracr*, *offset*, *ns*, and *dt* are read from the file. However, instead of writing the key values to the screen, the key values are directed to a new file, *test.txt*. We can see the contents of *test.txt* by using the **cat** command.

```
sugethw < myplot.su key=tracl,tracr,offset,ns,dt > test.txt
cat test.txt
```

Below are the contents of file *test.txt*. (Blank lines were removed from the file.)

```
tracl=1     tracr=1     offset=400     ns=64     dt=4000
tracl=2     tracr=2     offset=400     ns=64     dt=4000
tracl=3     tracr=3     offset=400     ns=64     dt=4000
tracl=4     tracr=4     offset=400     ns=64     dt=4000
tracl=5     tracr=5     offset=400     ns=64     dt=4000
tracl=6     tracr=6     offset=400     ns=64     dt=4000
tracl=7     tracr=7     offset=400     ns=64     dt=4000
tracl=8     tracr=8     offset=400     ns=64     dt=4000
tracl=9     tracr=9     offset=400     ns=64     dt=4000
tracl=10    tracr=10    offset=400     ns=64     dt=4000
tracl=11    tracr=11    offset=400     ns=64     dt=4000
tracl=12    tracr=12    offset=400     ns=64     dt=4000
tracl=13    tracr=13    offset=400     ns=64     dt=4000
tracl=14    tracr=14    offset=400     ns=64     dt=4000
tracl=15    tracr=15    offset=400     ns=64     dt=4000
tracl=16    tracr=16    offset=400     ns=64     dt=4000
tracl=17    tracr=17    offset=400     ns=64     dt=4000
tracl=18    tracr=18    offset=400     ns=64     dt=4000
tracl=19    tracr=19    offset=400     ns=64     dt=4000
tracl=20    tracr=20    offset=400     ns=64     dt=4000
tracl=21    tracr=21    offset=400     ns=64     dt=4000
tracl=22    tracr=22    offset=400     ns=64     dt=4000
tracl=23    tracr=23    offset=400     ns=64     dt=4000
tracl=24    tracr=24    offset=400     ns=64     dt=4000
tracl=25    tracr=25    offset=400     ns=64     dt=4000
tracl=26    tracr=26    offset=400     ns=64     dt=4000
tracl=27    tracr=27    offset=400     ns=64     dt=4000
tracl=28    tracr=28    offset=400     ns=64     dt=4000
tracl=29    tracr=29    offset=400     ns=64     dt=4000
tracl=30    tracr=30    offset=400     ns=64     dt=4000
tracl=31    tracr=31    offset=400     ns=64     dt=4000
tracl=32    tracr=32    offset=400     ns=64     dt=4000
```

As we suspected, the traces are numbered successively. As we knew, all traces have offset value 400.

3.3   *Real Seismic Shot Gathers*

Thanks to the Center for Wave Phenomena (CWP) at the Colorado School of Mines (CSM), we can download any of forty shot gathers collected from around the world. These are described in Yilmaz (1989) in Table 1-8 and pages 26-41, and Yilmaz (2001) in Table 1-13 and pages 67-80. You can download these files from:

```
ftp://ftp.cwp.mines.edu/pub/data/oz.original/
```

Be sure to read the README file on this site:

```
ftp://ftp.cwp.mines.edu/pub/data/oz.original/README
```

The following sections show how to download shot gather 3 from the CWP web site and test the gather's "endianness." Endianness refers to whether the downloaded binary file is big-endian (high byte) or little-endian (low byte). You don't have to understand endianness to understand the following sections. If you want to understand it, put "endianness" in a web search engine.

3.3.1 Get the Data

Go to the web site cited earlier and right-mouse click on *ozdata.3*. A pull-down menu appears. If you are using Netscape or Mozilla, click on "Save Link Target As..." If you are using Internet Explorer, click on "Save Target As..." If you are using a different browser, select a similar item from the right mouse pull-down menu. Save the file to your seismic data directory.

3.3.2 Test the Data

3.3.2.1 Big-Endian (High Byte) Machine: Solaris

At Michigan Tech, I (Forel) use a Solaris computer. According to the README file cited earlier, Solaris is a big-endian machine, the same endianness as the data on the CWP web site. Below, I test the file by using **surange**. (We first saw **surange** in Section 3.2). My computer prompt is "`/home/forel/suscripts/data/>`."

```
1   /home/forel/suscripts/data/> surange < ozdata.3
2   24 traces:
3    tracl=(1,24)  tracr=(1,24)  fldr=10003 tracf=(1,24)  cdp=(3,26)
4    cdpt=1 trid=1 nvs=1 nhs=1 duse=1
5    scalel=1 scalco=1 counit=1 delrt=4 muts=4
6    ns=1550 dt=4000
7   /home/forel/suscripts/data/> mv  ozdata.3  oz03.su
8   /home/forel/suscripts/data/>
```

In this test:

- In line 1, I input the data to **surange**.
- Lines 2-6 are the output from **surange**. Program **surange** did not have any problem reading the file.
- In line 7, I rename the file so it has our naming convention.

3.3.2.2 Little-Endian (Low Byte) Machine: PC

At home, I (Forel) use a Linux computer. According to the README file cited earlier, my PC is a little-endian machine, NOT the same endianness as the data on the CWP web site. Again, I test the file by using **surange**. My computer prompt is "`/home/david/suscripts/data/$`."

```
1   /home/david/suscripts/data/$ surange < ozdata.3
2
3   surange: fgettr.c: on trace #2 number of samples in header (212)
4    differs from number for first trace (32520)
5   /home/david/suscripts/data/$ suswapbytes < ozdata.3 | surange
6   24 traces:
7    tracl=(1,24)  tracr=(1,24)  fldr=10003 tracf=(1,24)  cdp=(3,26)
```

```
 8     cdpt=1 trid=1 nvs=1 nhs=1 duse=1
 9     scalel=1 scalco=1 counit=1 delrt=4 muts=4
10     ns=1550 dt=4000
11     /home/david/suscripts/data/$ suswapbytes < ozdata.3 > oz03.su
12     /home/david/suscripts/data/$ surange < oz03.su
13     24 traces:
14     tracl=(1,24)  tracr=(1,24)  fldr=10003 tracf=(1,24)  cdp=(3,26)
15     cdpt=1 trid=1 nvs=1 nhs=1 duse=1
16     scalel=1 scalco=1 counit=1 delrt=4 muts=4
17     ns=1550 dt=4000
18     /home/david/suscripts/data/$
```

In this test:

- In line 1, I input the file to **surange**.
- Lines 3-4 show that **surange** cannot read the file.
- In line 5, because I suspect the problem is the endianness of the file, I use **suswapbytes** to change the file's endianness, then I pipe the altered file to **surange**.
- Lines 6-10 show that **surange** can read the altered file.
- In line 11, I use **suswapbytes** to change the endianness of the input file, then I redirect the output to a file with our naming convention.
- In line 12, I test the new file with **surange**.
- Lines 13-17 show that the new file can be read by **surange**.

3.4   *Program suxwigb and Option perc*

Download file *ozdata.3* from the CWP ftp site. Use our file naming convention – change the file name to *oz03.su* by using the Unix **mv** command:

```
mv  ozdata.3  oz03.su
```

Look at the data by using the **suxwigb** program:

```
suxwigb  <  oz03.su  perc=95
```

The *perc* option is useful for examining real data. (Synthetic data are usually evenly balanced.) The *perc* option stands for "percentile for determining clip." Its practical use is to gain all traces by approximately the same amount. Try various value of *perc*, perhaps from 80 to 99. If *perc* is not specified, **suxwigb** uses the default value *perc=100*.

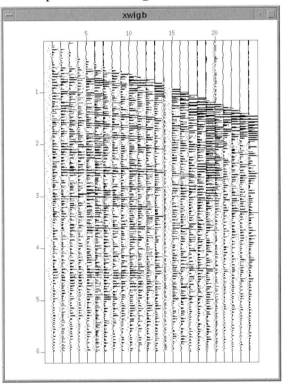

Figure 3.1: *Data set oz03.su displayed with perc=95.*

3.5   *Second Look at Trace Headers (keys): sukeyword*

Below is the **surange** output of file *oz03.su*.

```
surange  <  oz03.su
24 traces:
 tracl=(1,24)  tracr=(1,24)  fldr=10003 tracf=(1,24)  cdp=(3,26)
 cdpt=1 trid=1 nvs=1 nhs=1 duse=1
 scalel=1 scalco=1 counit=1 delrt=4 muts=4
 ns=1550 dt=4000
```

Again, program **surange** is a rapid way to see the smallest/largest values of the non-zero value keys in a data set. (If all the values of a key are zero, **surange** does not print the key.) These traces have different *cdp* values, and there are no offset values. (You can use **sugethw** to confirm this.)

Thought points:

- We do not expect *dt*, the sample interval key, to have more than one value.
- We expect *tracf*, the trace number key, to exhibit many values.

For a complete list of SU keys, use program **sukeyword**:

```
sukeyword  -o
```

This program prints a document, one page at a time to the screen. Press the space bar repeatedly to scroll to the next pages of the listing. The "–o" flag tells the program that prints the contents of file *segy.h* to start the screen print with the first key definition.

Alternatively, you can send the listing to a file, like this:

```
sukeyword  -o  >  sukeys.txt
```

Now you can either open file *sukeys.txt* (I made up the name.) in a text editor for convenient reading or you can use **more** to scroll through the file:

```
more  sukeys.txt
```

If you know the name of a key and want its definition, use the following command:

```
sukeyword  [ ]
```

For example, when you type:

```
sukeyword  dt
```

the screen print of file *segy.h* opens and quickly scrolls to put the definition of *dt* just a few lines below the top of the screen. Try it!

3.6   Third Look at Trace Headers (keys): suxedit

Program **suxedit** is another way to learn which keys are in a data set. Program **suxedit** is interactive. When you enter:

```
suxedit  <  oz03.su
```

the keys and their values are immediately listed for the first trace that is read. (**Note:** Sometimes traces are numbered in reverse order. When that happens, the first trace read is the one with the largest trace number.) You can then go forward or backward through the data set to see key values of other traces. When you are in **suxedit**, enter "?" to get a list of options.

3.7   Trace Header Manipulation: Constant and Regularly Varying Values

Let's go beyond trace header viewing. Let's modify some trace headers in *oz03.su*.

We saw in Section 3.5 that the traces in *oz03.su* have different cdp values, and all offset values are zero. To use some of our processes (like velocity analysis), we want the gather to appear to be a CMP (CDP). So, all the traces in the gather must have the same value in key *cdp*.

We can solve the CDP problem with a one-line command that uses program **sushw**: set one or more header words. This command replaces all *cdp* values with a single value and creates a new seismic file *oz03h1.su*.

```
sushw  <  oz03.su  key=cdp  a=3  >  oz03h1.su
```

See the selfdoc for **sushw**.

Now let's consider how to put offset values into the trace headers. First, we want to learn what the offset values should be. Yilmaz describes this shot record in Table 1-8 of Seismic Data Processing (1987) and in Table 1-13 of Seismic Data Analysis (2001):

Table 3.2: Shot gather 3

Number of Samples per Trace	Sampling Interval (ms)	Number of Traces	Trace Interval (ft)	Inner Offset (ft)
1500	4	24	340	340

We see from Figure 3.1 that offset increases with trace number (header *tracl*, *tracr*, or *tracf*.)

The following one-line command generates the appropriate *offset* values and creates a new seismic file *oz03h2.su*.

```
sushw < oz03h1.su key=offset a=340 b=340 j=24 > oz03h2.su
```

Let's use **surange** to examine the trace headers:

```
surange   <   oz03h2.su
```

The **surange** output is:

```
24 traces:
 tracl=(1,24)   tracr=(1,24)   fldr=10003 tracf=(1,24)   cdp=3
 cdpt=1 trid=1 nvs=1 nhs=1 duse=1
 offset=(340,8160)   scalel=1 scalco=1 counit=1 delrt=4
 muts=4 ns=1550 dt=4000
```

Let's look at some of the trace header information:

1. *tracl, tracr, tracf*:   Traces are numbered 1 to 24.
2. *cdp*:   Trace cdp numbers are the single value "3."
3. *offset*: The offsets range from 340 to 8160 (24 traces x 340 ft = 8160 ft).
4. *dt*:   The sample interval is 4 ms.
5. *delrt*: The time delay to the first sample is 4 ms (one sample).

Note: You can use **sugethw** to confirm these key values.

3.8   Trace Header Manipulation 2: Reverse Numbering

Download file *ozdata.30* from the CWP ftp site. Use our file naming convention – change the file name to *oz30.su* by using the Unix **mv** command:

```
mv  ozdata.30  oz30.su
```

Look at the data by using the **suxwigb** program:

```
suxwigb  <  oz30.su  perc=85
```

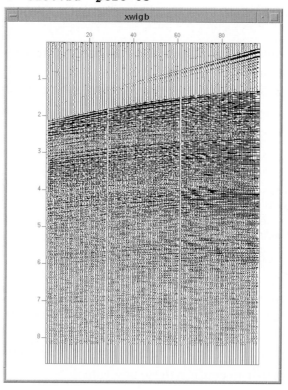

Figure 3.2: *Data set oz30.su displayed with perc=85.*

Below is the **surange** output of this file.

```
        surange  <  oz30.su
96 traces:
 tracl=(1,96)  tracr=(1,96)  fldr=10030 tracf=(1,96)  cdp=(30,125)
 cdpt=1 trid=1 nvs=1 nhs=1 duse=1
 scalel=1 scalco=1 counit=1 delrt=4 muts=4
 ns=2175 dt=4000
```

Again, traces have different cdp values, and there are no offset values. And, to use some of our processes we want the gather to appear to be a CMP (CDP). So, all the traces in the gather must have the same value in key *cdp*.

We already know how to replace all *cdp* values with a single value, but we show it again for fun.

```
        sushw  <  oz30.su  key=cdp  a=30  >  oz30h1.su
```

Before we put offset values into the trace headers, we want to learn what the offset values should be. Again, we refer to Table 1-8 of Seismic Data Processing (Yilmaz, 1987) or Table 1-13 of Seismic Data Analysis (Yilmaz, 2001):

Table 3.3: Shot gather 30

Number of Samples per Trace	Sampling Interval (ms)	Number of Traces	Trace Interval (m)	Inner Offset (m)
2125	4	96	25	230

We see from Figure 3.2 that offset *decreases* with increasing trace number. Before we use **sushw**, we have to calculate the offset of the first trace, the farthest offset trace:

$$\text{farthest offset} = 230 + (95 * 25) = 2605$$

The following one-line command generates the appropriate *offset* values and creates a new seismic file *oz30h2.su*.

```
sushw < oz30h1.su key=offset a=2605 b=-25 j=96 > oz30h2.su
```

Notice that *a*, the first trace, is assigned the value of the farthest offset and we use the negative trace interval in *b* to count down from the farthest offset.

Let's use **surange** to examine the trace headers:

```
surange  <  oz30h2.su
```

The **surange** output is:

```
96 traces:
 tracl=(1,96)  tracr=(1,96)  fldr=10030 tracf=(1,96)  cdp=30
 cdpt=1 trid=1 nvs=1 nhs=1 duse=1
 offset=(230,2605)  scalel=1 scalco=1 counit=1 delrt=4
 muts=4 ns=2175 dt=4000
```

Let's look at some of the trace header information:

1. *tracl, tracr, tracf*:   Traces are numbered 1 to 96.
2. *cdp*:   Trace cdp numbers are the single value "30".
3. *offset*:   The offsets range from 230 to 2605.
4. *dt*:   The sample interval is 4 ms.
5. *delrt*:   The time delay to the first sample is 4 ms (one sample).

Note: You can use **sugethw** to confirm these key values.

There are richer (more complex) ways to use **sushw** than we have shown. We leave that exploration to you.

3.9 *Windowing Data*

The steps below give the details of how the data/figures below were obtained.

Step 1   We "select" or "window" a single trace, 60, from original file *oz30.su*.

```
suwind < oz30.su key=tracl min=60 max=60 > oz3060.su
```

Step 2   Now trace 60 is the only data in new file *oz3060.su* (the left part of Figure 3.3). Display the file as:

```
suxwigb < oz3060.su &
```

Step 3   From the file with single trace 60, we window the time interval of 1 to 3 seconds.

```
suwind < oz3060.su tmin=1 tmax=3 > oz3060t.su
```

Step 4   We create the right wiggle trace display to show *oz3060t.su*.

```
suxwigb < oz3060t.su &
```

Step 5   We list the files to see relative sizes.

```
ls -1F oz30*.su
858240   Dec  7  16:58  oz30.su
  8940   Dec  7  17:09  oz3060.su
  2244   Dec  7  17:13  oz3060t.su
```

Step 6   We use **surange** to examine trace headers of the single trace.

```
surange < oz3060.su
```

```
1 traces:
tracl=60 tracr=60 fldr=10030 tracf=60 cdp=89
cdpt=1 trid=1 nvs=1 nhs=1 duse=1
scalel=1 scalco=1 counit=1 delrt=4 muts=4
ns=2175 dt=4000
```

Step 7   We use **surange** to examine trace headers of the time-windowed trace..

```
surange < oz3060t.su
```

```
1 traces:
tracl=60 tracr=60 fldr=10030 tracf=60 cdp=89
cdpt=1 trid=1 nvs=1 nhs=1 duse=1
scalel=1 scalco=1 counit=1 delrt=1000 muts=4
ns=501 dt=4000
```

Step 8   Examine both **surange** runs. Notice that header *delrt* (delay recording time in milliseconds) retains the old time annotation. That is, although *oz3060t.su* is a new file, header *delrt* tells the display program how to annotate the windowed segment.

Thought 1: In Step 1, if the input file contains many gathers, this step would create a file of every trace 60 from all the gathers.

Thought 2: In Step 3, if the input gather has many traces, this step would create a file of every trace time-limited from 1 to 3 seconds.

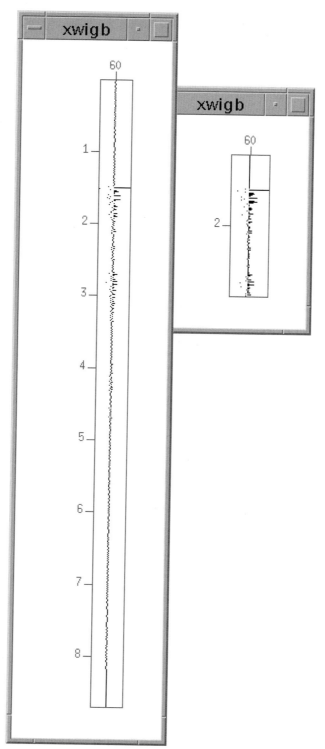

Figure 3.3: *Left: Trace 60 from oz30.su. Right: Trace 60 windowed from 1 to 3 seconds.*

4. Three Simple Models

4.1 Introduction

This chapter shows you how to create 2-D geologic models, create synthetic shot gathers from the models, and examine the shot gathers for quality control (QC the gathers). Chapter 5 shows you how to use the models to "acquire" (create synthetic) seismic data sets.

Models developed in this chapter have layers that are homogeneous and isotropic. Each layer has a single acoustic (P-wave) velocity.

This chapter and the next are computer-intensive in two ways: (1) the scripts are complex and (2) it will probably take your computer two hours to most of a day to generate the shot gathers. If you want to skip this complexity, skim these two chapters to familiarize yourself with the geologic models, then simply use the synthetic data set generated from Model 4 (Chapter 6) that accompanies this Primer. Seismic data generated from Model 4 are processed in Chapters 7, 8, and 9.

4.2 Model 1: Earth Model – Five Layers

Our first model consists of five homogenous, isotropic layers. In the x-direction, the model goes from zero to six kilometers. In the z-direction, the model goes from zero to two kilometers. In Figure 4.1, as with all the model images in this Primer, the model is drawn 1:1 (the horizontal units have the same length as the vertical units; the model is not stretched or squeezed). It is important to see a model 1:1 to assess the complexity of source-receiver raypaths.

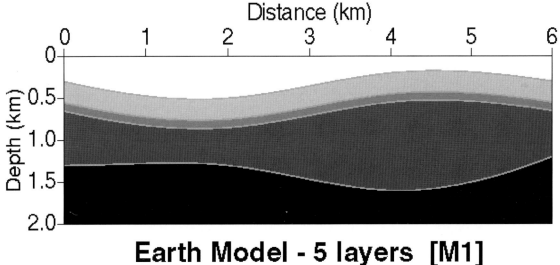

Figure 4.1: *Model 1: Five homogeneous, isotropic layers.*

Let's examine the script that created Model 1 and Figure 4.1. The numbers on the left are added for discussion; they are not part of the script.

```
1   #! /bin/sh
2   # File: model1.sh
3
4   # Set messages on
5   set -x
```

```
 6
 7   # Experiment Number
 8   num=1
 9
10   # Name output binary model file
11   modfile=model${num}.dat
12
13   # Name output encapsulated Postscript image file
14   psfile=model${num}.eps
15
16   # Remove previous .eps file
17   rm -f $psfile
18
19   trimodel xmin=0 xmax=6 zmin=0 zmax=2 \
20   1 xedge=0,6 \
21     zedge=0,0 \
22     sedge=0,0 \
23   2 xedge=0,2,4,6 \
24     zedge=0.30,0.50,0.20,0.30 \
25     sedge=0,0,0,0 \
26   3 xedge=0,2,4,6 \
27     zedge=0.55,0.75,0.45,0.55 \
28     sedge=0,0,0,0 \
29   4 xedge=0,2,4,6 \
30     zedge=0.65,0.85,0.55,0.65 \
31     sedge=0,0,0,0 \
32   5 xedge=0,2,4,6 \
33     zedge=1.30,1.30,1.60,1.20 \
34     sedge=0,0,0,0 \
35   6 xedge=0,6 \
36     zedge=2,2 \
37     sedge=0,0 \
38    kedge=1,2,3,4,5,6 \
39    sfill=0.1,0.1,0,0,0.44,0,0 \
40    sfill=0.1,0.4,0,0,0.40,0,0 \
41    sfill=0.1,0.6,0,0,0.35,0,0 \
42    sfill=0.1,1.0,0,0,0.30,0,0 \
43    sfill=0.1,1.5,0,0,0.25,0,0 > $modfile
44   ##        x,z
45
46   # Create a Postscript file of the model
47   #    Set gtri=1.0 to see sloth triangle edges
48   spsplot < $modfile > $psfile \
49           gedge=0.5 gtri=2.0 gmin=0 gmax=1 \
50           title="Earth Model - 5 layers  [M${num}]" \
51           labelz="Depth (km)" labelx="Distance (km)" \
52           dxnum=1.0 dznum=0.5 wbox=6 hbox=2
53
54   # Exit politely from shell
55   exit
56
```

This script can be divided into sets:

- System: Line 1 invokes the shell, line 5 turns on messages, and line 55 exits the shell.

- Variables:

  o Line 8 lets us vary the number of each run as we perfect the script. This number becomes part of output file names. Examples: 1, 101, 1a.

  o Line 11 assigns a name to the output binary model file that is used on line 43. **The output binary model file is the prime reason for this script.** Line 11 is an example of using curly brackets { } to concatenate text. Line 11 does not need this level of clarity, but this is a good place to introduce the idea. The curly brackets are necessary when we put text after a variable:

```
model${num}binary.dat
```

It is also good practice when using two variables next to each other:

```
model${variable1}${variable2}.dat
```

- o Line 14 assigns a name to the output .eps image file that is used on line 48.

- Bookkeeping: Line 17 removes a previous image file. This line is optional. On some systems, the program crashes if a .eps file with the same name already exists. Usually files are overwritten!

- Program **trimodel**: Lines 19-43 create the model. Notice that line 44 is a comment. We use line 44 to remind us that the first two entries of line *sfill* are x-z values. The very important *sfill* parameter is discussed below.

- Program **spsplot**: Lines 48-52 create the .eps image file.

Let's examine program **trimodel**. Program **trimodel** fills the model with triangles of $(1/velocity)^2$. While $(1/velocity)$ is called "slowness," $(1/velocity)^2$ is called "sloth." A sloth is a slow-moving tree-dwelling mammal found in Central and South America. Program **trimodel**'s use here can be divided into five parts:

1. Line 19 defines the model dimensions.

2. The six sets of *xedge*, *zedge*, and *sedge* triplets define layer boundaries and velocity gradients. There is a requirement that each set of triplets have the same number of values. In other words, the triplet for layer 1 that defines the top of the model (a straight line) has (only needs) two values and each triplet (*xedge*, *zedge*, *sedge*), therefore, contains two values. On the other hand, the triplet for layer 2 that defines a curved surface has more than two values.

```
20    1 xedge=0,6 \
21      zedge=0,0 \
22      sedge=0,0 \

23    2 xedge=0,2,4,6 \
24      zedge=0.30,0.50,0.20,0.30 \
25      sedge=0,0,0,0 \
```

A triplet consists of an *xedge*, *zedge* pair that is an interface control point and an *sedge* value that is the velocity gradient at the control point.

Line *sedge* has only zeros because all layers are isotropic and homogeneous. Line *sedge* would have non-zero values if we want a layer to have velocity gradients.

3. In this model that has six boundaries, there are five simple layers. Therefore, there are five *sfill* lines, lines 39-43. These lines are written for isotropic, homogenous layers, which is why most of the values are zero.

Table 4.1: Line *sfill* variables

x	z	x0	z0	s00	ds/dx	ds/dz

Each x-z pair of *sfill* is a point in a layer. Each *sfill* line describes the sloth value that fills the layer.

Table 4.2: Values of *sfill* in Model 1

x (m)	z (m)	s00 (sloth)	velocity (1/√sloth) (m/s)
0.1	0.1	0.44	1508
0.1	0.4	0.40	1581
0.1	0.6	0.35	1690
0.1	1.0	0.30	1826
0.1	1.5	0.25	2000

These P-wave velocity values are not very realistic (except the first), but they make a nice grey-scale image in the .eps file.

4. Line 38, line *kedge,* is necessary for any acquisition program that uses this model. **An interface that is not listed here will not be seen by a later acquisition script.**

5. Line 43 has the variable name of the output binary model file.

Program **spsplot** plots a triangulated sloth function as an encapsulated Postscript file. So, we see that **spsplot** is designed to accompany **trimodel**! If you want to see the triangles that are used to build the model, change *gtri=2.0* to *gtri=1.0.*

On line 52, *wbox=6* and *hbox=2* ensure that the .eps image has the same dimensions as the actual model (line 19). It is important for us to specify the same width and height in different programs within the same script so images created by different programs exactly overlap. You will see this in scripts in the second half of this chapter.

Also on line 52, *dxnum* and *dznum* specify the annotation increments in the respective directions. These parameters are important in later ray trace and wavefront scripts.

Below are 60-trace split-spread shot gathers acquired from source positions x=2.0, x=2.6, x=3.3, and x=3.95. Acquisition of these data is explained in Chapter 5.

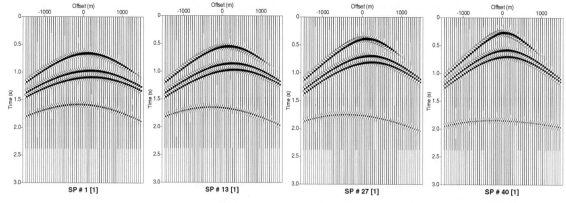

Figure 4.2: *Left to right: Shot 1, Shot 13, Shot 27, Shot 40 from Model 1.*

Remember, you can use the self-documentation to learn more about any SU program by entering just the program name on a line (Section 1.11).

4.3   Model 2: Earth Model – High Velocity Intrusion

Our second model consists of three homogenous, isotropic layers with a high velocity intrusion. In the x-direction, the model goes from zero to six kilometers. In the z-direction, the model goes from zero to two kilometers.

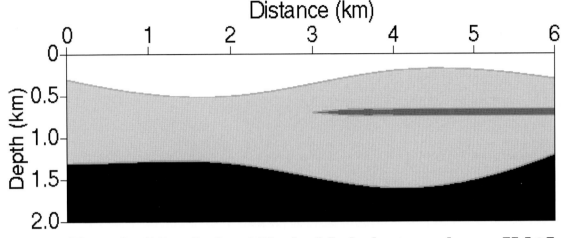

Figure 4.3: *Model 2: Three homogeneous layers with a high-velocity intrusion.*

As you can see in Figure 4.3, this model has four layers. But the script for Model 2 (below), like the script for Model 1, has six interfaces. The important difference here is that interface 3 and interface 4 share the same first x-z point. Because interfaces 3 and 4 "pinch" closed at x=3 and because they both extend to the right edge of the model, the "sloth" fill of layer 2 flows around interfaces 3 and 4. Therefore, only four *sfill* lines are needed.

```
 1   #! /bin/sh
 2   # File: model2.sh
 3
 4   # Set messages on
 5   set -x
 6
 7   # Experiment Number
 8   num=2
 9
10   # Name output binary model file
11   modfile=model${num}.dat
12
13   # Name output encapsulated Postscript image file
14   psfile=model${num}.eps
15
16   # Remove previous .eps file
17   rm -f $psfile
18
19   trimodel xmin=0 xmax=6 zmin=0 zmax=2 \
20   1 xedge=0,6 \
21     zedge=0,0 \
22     sedge=0,0 \
23   2 xedge=0,2,4,6 \
24     zedge=0.3,0.5,0.2,0.3 \
25     sedge=0,0,0,0 \
26   3 xedge=3,3.5,4,6 \
```

```
27    zedge=0.7,0.66,0.66,0.66 \
28    sedge=0,0,0,0 \
29  4 xedge=3,3.5,4,6 \
30    zedge=0.7,0.74,0.74,0.74 \
31    sedge=0,0,0,0 \
32  5 xedge=0,2,4,6 \
33    zedge=1.3,1.3,1.6,1.2 \
34    sedge=0,0,0,0 \
35  6 xedge=0,6 \
36    zedge=2,2 \
37    sedge=0,0 \
38   kedge=1,2,3,4,5,6 \
39   sfill=1,0.1,0,0,0.44,0,0 \
40   sfill=1,0.7,0,0,0.40,0,0 \
41   sfill=4,0.7,0,0,0.30,0,0 \
42   sfill=1,1.5,0,0,0.20,0,0 > $modfile
43  ##     x,z
44
45  # Create a Postscript file of the model
46  spsplot < $modfile > $psfile \
47          gedge=0.5 gtri=2.0 gmin=0 gmax=1 \
48          title="Earth Model - High Vel. Intrusion  [M${num}]" \
49          labelz="Depth (km)" labelx="Distance (km)" \
50          dxnum=1.0 dznum=0.5 wbox=6 hbox=2
51
52  # Exit politely from shell
53  exit
54
```

Table 4.3: Values of *sfill* in Model 2

x (m)	z (m)	s00 (sloth)	velocity (1/√sloth) (km/s)
1	0.1	0.44	1508
1	0.7	0.40	1581
4	0.7	0.30	1826
1	1.5	0.20	2236

In this isotropic, homogenous model, *sfill* supplies (fills) a single sloth value for a layer when an x-z point is specified in the appropriate layer.

Below are 60-trace split-spread shot gathers acquired from source positions x=2.0, x=2.6, x=3.3, and 3.95. Acquisition of these data is explained in Chapter 5.

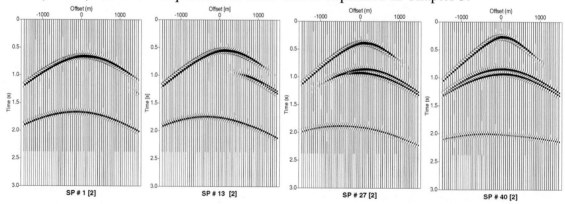

Figure 4.4: *Left to right: Shot 1, Shot 13, Shot 27, Shot 40 from Model 2.*

4.4   Model 3: Earth Model – Diffractor

Our third model consists of three homogenous, isotropic layers with an imbedded high velocity diffractor. In the x-direction, the model goes from -1 to +5 kilometers. In the z-direction, the model goes from 0 to 2 kilometers. See line 19.

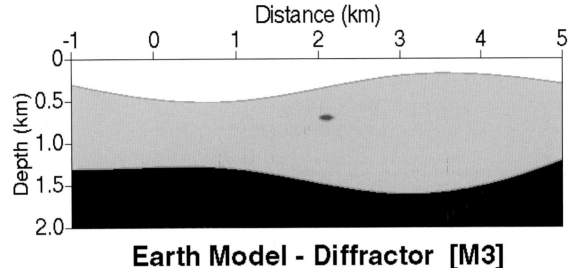

Figure 4.5: *Model 3: Three homogeneous layers with a diffracting lens.*

Like the previous models, this model has six interfaces. This time, interfaces 3 and 4 pinch closed on both sides; they extend for only 0.2 km. Again, the "sloth" fill of layer 2 flows around interfaces 3 and 4. Therefore, only four *sfill* lines are needed.

```
1   #! /bin/sh
2   # File: model3.sh
3
4   # Set messages on
5   set -x
6
7   # Experiment Number
8   num=3
9
10  # Name output binary model file
11  modfile=model${num}.dat
12
13  # Name output encapsulated Postscript image file
14  psfile=model${num}.eps
15
16  # Remove previous .eps file
17  rm -f $psfile
18
19  trimodel xmin=-1 xmax=5 zmin=0 zmax=2 \
20  1 xedge=-1,5 \
21    zedge=0,0 \
22    sedge=0,0 \
23  2 xedge=-1.,1.0,3.0,5.0 \
24    zedge=0.3,0.5,0.2,0.3 \
25    sedge=0,0,0,0 \
26  3 xedge=2.00,2.10,2.20 \
27    zedge=0.70,0.66,0.70 \
28    sedge=0,0,0 \
29  4 xedge=2.00,2.10,2.20 \
30    zedge=0.70,0.74,0.70 \
31    sedge=0,0,0 \
```

```
32  5 xedge=-1.,1.0,3.0,5.0 \
33    zedge=1.3,1.3,1.6,1.2 \
34    sedge=0,0,0,0 \
35  6 xedge=-1,5 \
36    zedge=2,2 \
37    sedge=0,0 \
38   kedge=1,2,3,4,5,6 \
39   sfill=1.0,0.1,0,0,0.44,0,0 \
40   sfill=1.0,0.6,0,0,0.40,0,0 \
41   sfill=2.1,0.7,0,0,0.30,0,0 \
42   sfill=1.0,1.5,0,0,0.20,0,0 > $modfile
43  ##       x,z
44
45  # Create a Postscript file of the model
46  spsplot < $modfile > $psfile \
47         gedge=0.5 gtri=2.0 gmin=0 gmax=1 \
48         title="Earth Model - Diffractor   [M${num}]" \
49         labelz="Depth (km)" labelx="Distance (km)" \
50         dxnum=1.0 dznum=0.5 wbox=6 hbox=2
51
52  # Exit politely from shell
53  exit
54
```

Table 4.4: Values of *sfill* in Model 3

x (m)	z (m)	s00 (sloth)	velocity (1/√sloth) (km/s)
1.0	0.1	0.44	1508
1.0	0.6	0.40	1581
2.1	0.7	0.30	1826
1.0	1.5	0.20	2236

The .eps image is still made 1:1 because in line 50, the box dimensions are six units wide by two units high.

Below are 60-trace split-spread shot gathers acquired from source positions x=1.0, x=1.6, x=2.3, and 2.95. Acquisition of these data is explained in Chapter 5.

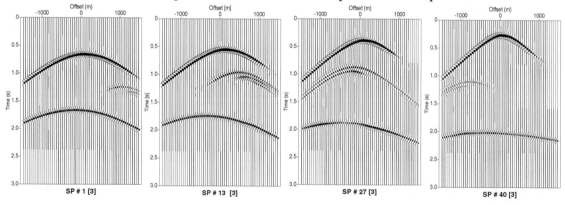

Figure 4.6: *Left to right: Shot 1, Shot 13, Shot 27, Shot 40 from Model 3.*

4.5 Introduction to Ray Tracing and Wavefronts

The following scripts trace rays and show wavefronts through the models discussed earlier. These images can help us select parameter values for acquiring full seismic data sets with the models. Mostly, the images are just fun to see.

The scripts are adapted from the examples in the SU demonstration directories *demos/Synthetic/Tri/Models* and *demos/Synthetic/Tri/Rays*. Our scripts create .eps images, then overlay them. The scripts are called "psmerge" because they are named for the **psmerge** program that merges .eps files. They all start with a .eps model file previously created.

In brief:

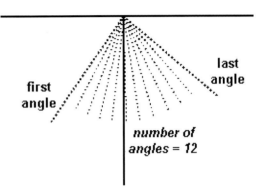

- We pick a location for a single shot, usually at the surface (z=0.0).

- We decide the number of rays that emanate from the source, defined by "number of angles" (*nangle*).

- We decide the angle of the fan of rays that emanate from the source, defined by "first angle" (*fangle*) and "last angle" (*langle*).

In the example above, *nangle=12*, *fangle=-30*, and *langle=45*. The larger the value of *nangle*, the longer it takes to produce the image. However, a complicated model might require many rays to ensure that some will reflect to surface receivers.

- We also decide whether the rays transmit (refract) through a layer or reflect from it. For example:
    - *refseq=2,0*   means the rays refract through interface 2
    - *refseq=5,1*   means the rays reflect at interface 5

Because .eps files are merged, they should have the same height and width. Also, the z-axis label and x-axis label of ray and wavefront images either should match the labels of the model image or the labels should not be generated. In the scripts below, labels are not generated for ray and wavefront images since the labels are already on the model image.

Axes labels **are** specified for ray and wavefront images because the default values might not match the values generated for the model image. When the same values are specified for each image, they overlap!

The main program, **triray**, generates the ray and wavefront information. It is up to us whether we save this information. Program **psgraph** is called once to map the ray information to a .eps image; it is called a separate time to map the wavefront information to a separate .eps image. Program **psmerge** is used once to merge however many .eps images were created.

4.6 Model 1 Ray Tracing: psmerge1a.sh

Using Model 1, a source at x=4.5, z=0.0 (line 25) generates 10 rays (*nangle*) from angles -45° to -10° (*fangle, langle*) (line 20). In addition to the final .eps merge file, only ray files are output (lines 14-16).

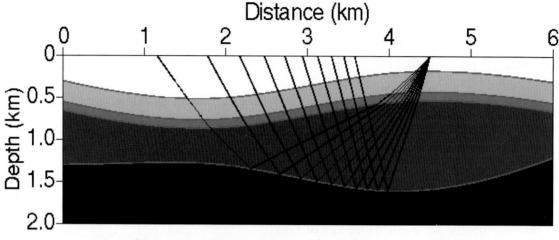

Figure 4.7: *Rays generated through Model 1.*

We used the *refseq* parameter to specify that the rays should refract through interfaces 2, 3, and 4, but should reflect from interface 5 (line 27). Remember that the top of the model is interface 1 and the bottom is interface 6. The input binary model file (line 11) carries sloth information for the model.

In the model file (Section 4.2), x-z increments are specified in line 52. To override the default values of **psgraph**, the x-z increments are again specified in line 33. Unfortunately, the variables in the two programs do not have the same names.

Table 4.5: z-x increment annotation parameters

program	z-increment	x-increment
trimodel	dznum	dxnum
psgraph	d1num	d2num

While *hbox=2.0* and *wbox=6.0* ensure that the model and ray images have the same dimensions (line 31 in this script and line 52 in the model script), **psgraph** still must be told the x and z extents of the image (line 32).

```
 1   #! /bin/sh
 2   # File: psmerge1a.sh
 3
 4   # Set messages on
 5   set -x
 6
 7   # Experiment number
 8   num=1
 9
10   # Input files
11   modelfile=model${num}.dat
12   modelpsfile=model${num}.eps
```

```
13
14    # Output files
15    rayendsfile=rayends${num}a.dat
16    rayfile=ray${num}a.dat
17    raypsfile=ray${num}a.eps
18    psmergefile=psmerge${num}a.eps
19
20    # Assign values to variables
21    nangle=10 fangle=-45 langle=-15 nxz=301
22
23    # Shoot the rays
24    triray < $modelfile > $rayendsfile rayfile=$rayfile \
25          nangle=$nangle fangle=$fangle langle=$langle \
26          xs=4.5 zs=0.0 nxz=$nxz \
27          refseq=2,0 refseq=3,0 refseq=4,0 refseq=5,1
28
29    # Plot the rays
30    psgraph < $rayfile > $raypsfile \
31          nplot=`cat outpar` n=$nxz hbox=2.0 wbox=6.0 \
32          x1beg=0.0 x1end=2.0 x2beg=0 x2end=6 \
33          d1num=0.5 d2num=1.0 style=seismic linegray=0
34
35    # Merge model + rays
36    psmerge in=$modelpsfile in=$raypsfile > $psmergefile
37
38    # Exit politely from shell
39    exit
40
```

The rays are black because *linegray=0* in psgraph (line 32).

This script (*psmerge1a.sh*) and the next script (*psmerge1b.sh*) both use files created during our earlier work with Model 1. The input files, lines 11 and 12 of *psmerge1a.sh* and *psmerge1b.sh*, both use variable *${num}* where *num=1*. The output files of *psmerge1a.sh* use variable *${num}a* to distinguish the output files of this script from the output files of the next script.

4.7   *Model 1 Wavefronts: psmerge1b.sh*

To add wavefronts to Model 1, several wavefront files are created (lines 17-19) and **psgraph** is used a second time to map the wavefronts. Wavefronts are not actually drawn on the image. Instead, marks are placed at appropriate places on the rays. To make the illusion of wavefronts, the number of rays is considerably increased from the previous script. Here, *nangle=70* (line 23).

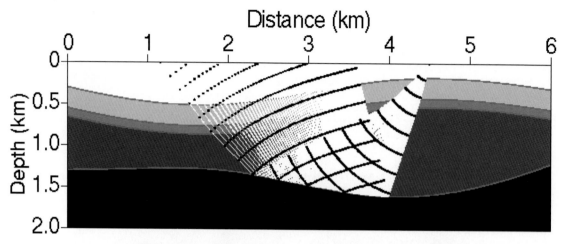

Earth Model - 5 layers [M1]

Figure 4.8: *Rays and wavefronts generated through Model 1.*

Parameter *nt* (line 24) specifies the number of time increments to draw on the advancing wavefront. Program **transp** (line 40) is necessary to transpose the wavefile image created by **triray**.

```
 1   #! /bin/sh
 2   # File: psmerge1b.sh
 3
 4   # Set messages on
 5   set -x
 6
 7   # Experiment number
 8   num=1
 9
10   # Input files
11   modelfile=model${num}.dat
12   modelpsfile=model${num}.eps
13
14   # Output files
15   rayendsfile=rayends${num}b.dat
16   rayfile=ray${num}b.dat
17   raypsfile=ray${num}b.eps
18   wavefile=wave${num}b.dat
19   wavetrans=wavetrans${num}b.dat
20   wavepsfile=wave${num}b.eps
21   psmergefile=psmerge${num}b.eps
22
23   # Assign values to variables
24   nangle=70 fangle=-45 langle=-15 nxz=301 nt=20
25
26   # Shoot the rays
27   triray < $modelfile > $rayendsfile \
28          rayfile=$rayfile wavefile=$wavefile \
29          nangle=$nangle fangle=$fangle langle=$langle \
30          nxz=$nxz nt=$nt xs=4.5 zs=0.0 \
31          refseq=2,0 refseq=3,0 refseq=4,0 refseq=5,1
32
33   # Plot the rays
34   psgraph < $rayfile > $raypsfile \
35          nplot=`cat outpar` n=$nxz hbox=2.0 wbox=6.0 \
36          x1beg=0.0 x1end=2.0 x2beg=0 x2end=6 \
37          d1num=0.5 d2num=1.0 style=seismic linegray=1
```

```
38
39   # Transpose the wavefile
40   transp < $wavefile > $wavetrans n1=$nt n2=$nangle nbpe=8
41
42   # Plot the wavefronts
43   psgraph < $wavetrans > $wavepsfile  \
44           linewidth=0.0 mark=8 marksize=2 \
45           nplot=$nt n=$nangle hbox=2.0 wbox=6.0 \
46           x1beg=0.0 x1end=2.0 x2beg=0.0 x2end=6.0 \
47           d1num=0.5 d2num=1.0 style=seismic linegray=0
48
49   # Merge model + rays + wavefronts
50   psmerge in=$modelpsfile in=$raypsfile in=$wavepsfile > $psmergefile &
51
52   # Exit politely from shell
53   exit
54
```

Each time **psgraph** is used, the image dimensions must be specified the same way (parameters *hbox, wbox, x1beg, x1end, x2beg, x2end, d1num, d2num*) to ensure that all merged images (line50) overlap exactly.

Now the rays are white due to *linegray=1* in line 37. The wavefronts are black due to *linegray=0* in line 47. Parameter *linegray* controls a gray continuum between 0 and 1.

4.8   Model 2 Ray Tracing: psmerge2a.sh

Using Model 2, a source at x=3.0, z=0.0 (line 25) generates 15 rays (*nangle*) from angles 0° to 50° (*fangle, langle*) (line 20). In addition to the final .eps merge file, only ray files are output (lines 14-16). We see some bending as rays pass through the intrusion.

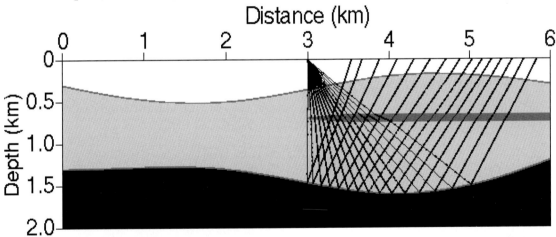

Figure 4.9: *Rays generated through Model 2.*

The *refseq* parameter specifies that the rays should refract through interfaces 2, 3, and 4, but should reflect from interface 5 (line 26). The input binary model file (line 10) carries sloth information for the model.

```
1   #! /bin/sh
2   # File: psmerge2a.sh
3
4   # Set messages on
5   set -x
6
```

```
 7  # Experiment number
 8  num=2
 9
10  # Input files
11  modelfile=model${num}.dat
12  modelpsfile=model${num}.eps
13
14  # Output files
15  rayendsfile=rayends${num}a.dat
16  rayfile=ray${num}a.dat
17  raypsfile=ray${num}a.eps
18  psmergefile=psmerge${num}a.eps
19
20  # Assign values to variables
21  nangle=15 fangle=0 langle=50 nxz=301
22
23  # Shoot the rays
24  triray < $modelfile > $rayendsfile rayfile=$rayfile \
25          nangle=$nangle fangle=$fangle langle=$langle \
26          xs=3.0 zs=0.0 nxz=$nxz \
27          refseq=2,0 refseq=3,0 refseq=4,0 refseq=5,1
28
29  # Plot the rays
30  psgraph < $rayfile > $raypsfile \
31          nplot=`cat outpar` n=$nxz hbox=2 wbox=6 \
32          x1beg=0 x1end=2 x2beg=0 x2end=6 \
33          d1num=0.5 d2num=1.0 style=seismic linegray=0
34
35  # Merge model + rays
36  psmerge in=$modelpsfile in=$raypsfile > $psmergefile &
37
38  # Exit politely from shell
39  exit
40
```

The rays are black because *linegray=0* in psgraph (line 32).

4.9   *Model 2 Wavefronts: psmerge2b.sh*

To add wavefronts to Model 2, the number of rays is considerably increased from the previous script. Here, *nangle=65* (line 24).

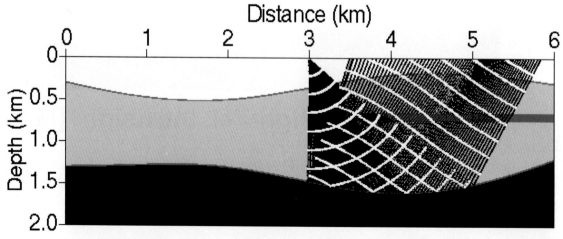

Figure 4.10: *Rays and wavefronts generated through Model 2.*

```
 1  #! /bin/sh
 2  # File: psmerge2b.sh
 3
 4  # Set messages on
 5  set -x
 6
 7  # Experiment number
 8  num=2
 9
10  # Input files
11  modelfile=model${num}.dat
12  modelpsfile=model${num}.eps
13
14  # Output files
15  rayendsfile=rayends${num}b.dat
16  rayfile=ray${num}b.dat
17  raypsfile=ray${num}b.eps
18  wavefile=wave${num}b.dat
19  wavetrans=wavetrans${num}b.dat
20  wavepsfile=wave${num}b.eps
21  psmergefile=psmerge${num}b.eps
22
23  # Assign values to variables
24  nangle=65 fangle=0 langle=50 nxz=301 nt=20
25
26  # Shoot the rays
27  triray < $modelfile > $rayendsfile \
28          rayfile=$rayfile wavefile=$wavefile \
29          nangle=$nangle fangle=$fangle langle=$langle \
30          nxz=$nxz nt=$nt xs=3.0 zs=0.0 \
31          refseq=2,0 refseq=3,0 refseq=4,0 refseq=5,1
32
33  # Plot the rays
34  psgraph < $rayfile > $raypsfile \
35          nplot=`cat outpar` n=$nxz hbox=2 wbox=6 \
36          x1beg=0 x1end=2 x2beg=0 x2end=6 \
37          d1num=0.5 d2num=1.0 style=seismic linegray=0
38
39  # Transpose the wavefile
40  transp < $wavefile > $wavetrans n1=$nt n2=$nangle nbpe=8
41
42  # Plot the wavefronts
43  psgraph < $wavetrans > $wavepsfile  \
44          linewidth=0.0 mark=8 marksize=2 \
45          nplot=$nt n=$nangle hbox=2.0 wbox=6.0 \
46          x1beg=0.0 x1end=2.0 x2beg=0.0 x2end=6.0 \
47          d1num=0.5 d2num=1.0 style=seismic linegray=1
48
49  # Merge model + rays + wavefronts
50  psmerge in=$modelpsfile in=$raypsfile in=$wavepsfile > $psmergefile &
51
52  # Exit politely from shell
53  exit
54
```

Now the rays are black due to *linegray=0* in line 36. The wavefronts are white due to *linegray=1* in line 47.

4.10 Model 3 Ray Tracing: psmerge3.sh

Using Model 3, a source at x=2.0, z=0.0 (line 26) generates 10 rays (*nangle*) from angles 0° to 25° (*fangle, langle*) (line 21). In addition to the final .eps merge file, only ray files are output (lines 15-17).

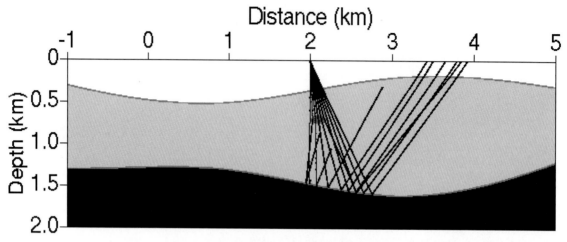

Earth Model - Diffractor [M3]

Figure 4.11: *Rays generated through Model 3. Program **triray** parameter nxz=301.*

The source was placed over the diffractor so rays shoot through and around it. In a real earth, raypaths would be quite scattered near the diffractor edges. Here, the edges are not treated realistically. Notice that only six of the ten rays reach the surface. In the first use of this script, **triray** parameter *nxz=301* (lines 21 and 26). To make all ten rays reach the surface, *nxz* was increased in increments of 50. All rays reached the surface with *nxz=701* (see Figure 4.12, below). It is interesting to observe that the default value of *nxz* is 101. It is probably unrealistic to set *nxz* so high (701), and probably a waste of computer time; although, without examining the source code and understanding exactly what role *nxz* plays, it is not easy to be certain. Nonetheless, through trial and error we obtained a reasonable solution..

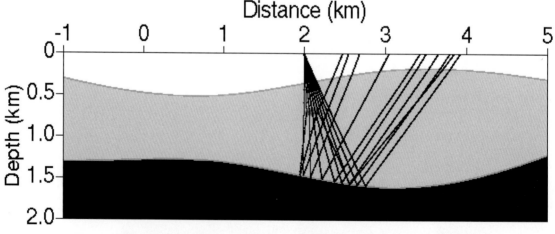

Earth Model - Diffractor [M3]

Figure 4.12: *Rays generated through Model 3. Program **triray** parameter nxz=701.*

While *hbox=2.0* and *wbox=6.0* ensure that the model and ray images have the same dimensions (line 31 in this script and line 50 in the model script), **psgraph** still must be told the x and z extents of the image (line 32). Notice that *x2beg=-1* and *x2end=5*.

```
 1  #! /bin/sh
 2  # File: psmerge3.sh
 3
 4  # Turn on messages
 5  set -x
 6
 7  # Experiment number
 8  num=3
 9
10  # Input files
11  modelfile=model${num}.dat
12  modelpsfile=model${num}.eps
13
14  # Output files
15  rayendsfile=rayends${num}.dat
16  rayfile=ray${num}.dat
17  raypsfile=ray${num}.eps
18  psmergefile=psmerge${num}.eps
19
20  # Assign values to variables
21  nangle=10 fangle=0 langle=+25 nxz=701
22
23  # Shoot the rays
24  triray < $modelfile > $rayendsfile rayfile=$rayfile \
25          nangle=$nangle fangle=$fangle langle=$langle \
26          xs=2.0 zs=0.0 nxz=$nxz \
27          refseq=2,0 refseq=3,0 refseq=4,0 refseq=5,1
28
29  # Plot the rays
30  psgraph < $rayfile > $raypsfile \
31          nplot=`cat outpar` n=$nxz hbox=2 wbox=6 \
32          x1beg=0 x1end=2 x2beg=-1 x2end=5 \
33          d1num=0.5 d2num=1.0 style=seismic linegray=0
34
35  # Merge model + rays
36  psmerge in=$modelpsfile in=$raypsfile > $psmergefile &
37
38  # Exit politely from shell
39  exit
40
```

5. Three Simple Models: Acquire 2-D Lines

5.1 Introduction

In the previous chapter, we developed three simple models and used them to generate single shot gathers and images of rays and wavefronts. In this chapter, we use the same models to acquired 2-D lines of seismic data.

5.2 Acquire 2-D Line from Model 1: acq1.sh

Script *acq1.sh* acquires seismic data over model *model1.dat*, generated by script *model1.sh* (Section 4.2). The survey layout:

1. 40 shots (line 28)
2. Shots are equally spaced at 50m intervals (line 32 and line 35)
3. 60 split-spread traces are recorded from each shot location (line 42)
4. Geophone spacing is 50m (line 46 and line 49)
5. Geophone offsets range from -1475 m to +1475 m (line 52)
6. Shot locations range from 2 km to 3.95 km (line 68)
7. Geophone locations range from 0.525 km to 5.425 km (line 68)

The final data set has 2400 traces (40 shots x 60 geophones per shot). The generation of this data set took about half an hour on a Sun UltraSPARC III with four processors.

```
 1   #! /bin/sh
 2   # acq1.sh
 3   # Set messages on
 4   ##set -x
 5
 6   # Assign values to variables
 7   num=1
 8   nangle=201 fangle=-65 langle=65
 9   nt=751 dt=0.004
10
11   # Name input model file
12   inmodel=model$num.dat
13
14   # Name output seismic file
15   outseis=seis$num.su
16
17   #==================================================
18   # Create the seismic traces with "triseis"
19   #    i-loop = 40 source positions
20   #    j-loop = 60 geophone positions (split-spread)
21   #             per shot position
22   #    k-loop = layers 2 through 5
23   #             (do not shoot layers 1 and 6)
24
25   echo " --Begin looping over triseis."
26
27   i=0
28   while [ "$i" -ne "40" ]
29   do
30
31     fs=`bc -l <<-END
32     $i * 0.05
33     END`
34     sx=`bc -l <<-END
35     $i * 50
```

```
36      END`
37      fldr=`bc -l <<-END
38      $i + 1
39      END`
40
41      j=0
42      while [ "$j" -ne "60" ]
43      do
44
45        fg=`bc -l <<-END
46        $i * 0.05 + $j *0.05
47        END`
48        gx=`bc -l <<-END
49        $i * 50 + $j * 50 - 1475
50        END`
51        offset=`bc -l <<-END
52        $j * 50 - 1475
53        END`
54        tracl=`bc -l <<-END
55        $i * 60 + $j + 1
56        END`
57        tracf=`bc -l <<-END
58        $j + 1
59        END`
60
61        echo " Sx=$sx   Gx=$gx    fldr=$fldr    Offset=$offset    tracl=$tracl\
62        fs=$fs     fg=$fg"
63
64        k=2
65        while [ "$k" -ne "6" ]
66        do
67
68          triseis < $inmodel  xs=2,3.95  xg=0.525,5.425  zs=0,0 zg=0,0 \
69                  nangle=$nangle fangle=$fangle langle=$langle \
70                  kreflect=$k krecord=1 fpeak=12 lscale=0.5 \
71                  ns=1 fs=$fs ng=1 fg=$fg nt=$nt dt=$dt |
72          suaddhead nt=$nt |
73          sushw key=dt,tracl,tracr,fldr,tracf,trid,offset,sx,gx \
74                  a=4000,$tracl,$tracl,$fldr,$tracf,1,$offset,$sx,$gx >> temp$k
75
76          k=`expr $k + 1`
77
78        done
79        j=`expr $j + 1`
80
81      done
82      i=`expr $i + 1`
83
84  done
85
86  echo " --End looping over triseis."
87
88  #================================================
89
90  # Sum contents of the "temp" files
91  echo " --Sum files."
92  susum temp2 temp3 > tempa
93  susum tempa temp4 > tempb
94  susum tempb temp5 > $outseis
95
96  # Remove temp files
97  echo " --Remove temp files."
98  rm -f temp*
```

```
 99
100  # Exit politely from shell script
101  echo " --Finished!"
102  exit
103
```

This script writes messages to the screen to help us track progress:

```
 25  echo " --Begin looping over triseis."
 61     echo " Sx=$sx   Gx=$gx   fldr=$fldr   Offset=$offset   tracl=$tracl\
 62     fs=$fs   fg=$fg"
 86  echo " --End looping over triseis."
 91  echo " --Sum files."
 97  echo " --Remove temp files."
101  echo " --Finished!"
```

Lines 61 and 62 write variable values to the screen. Notice that line 4 is commented out. We find that once a script is perfected, the messages from line 4 interfere with reading the values printed to the screen by lines 61 and 62.

Previously (Section 4.5), we discussed variables *nangle* (the number of rays or angles that emanate from the source), *fangle* (the first angle of the fan of rays that emanate from the source), and *langle* (the last angle of the fan of rays that emanate from the source). Here (and in the following acquisition scripts), these values are assigned on line 8.

Line 9 assigns the number of time samples (*nt*) and the time sample interval in seconds (*dt*).

Note: Check the unit of the time sample interval in each SU program. The unit varies.

The core of the script is the three `do-while` loops (Section 2.3.13) over **triseis**, **suaddhead**, and **sushw**. Each time these programs are used, **triseis** creates a seismic trace, **suaddhead** creates a trace header on the trace, and **sushw** writes values to the trace headers.

- The outermost i-loop goes over 40 source positions, 0 to 39 (lines 27 and 28).

- The j-loop goes over 60 geophone positions, 0 to 59 (lines 41 and 42).

- The innermost k-loop goes over the reflectors, 2 to 5 (lines 64 and 65).

Consider the innermost loop. The value of k signifies the reflector from which **triseis** records a reflection (*kreflect=$k*, line 70) and the *temp* file in which the trace is stored (*temp$k*, line 74.) Because k loops over reflectors 2 to 5, four *temp* files are created. (The top and base of the model are ignored.) Each *temp* data set has (in this case) 2400 traces. After all traces are created, lines 93 to 95 "sum" the four data sets, trace-for-trace.

The following pseudo-code of the `do-while` loops is presented to help us consider how the variables are used. The program in which the variable is used is printed in bold to the right on the assignment line. Variable k is underlined to remind us of its use as discussed in the preceding paragraph.

Variable *fs* (line E) is the source location relative to the start of the source surface. The source surface is defined in line Y as variable *xs*. As i increments, the source moves across the model. Both *fs* and *sx* are the source position. Variable *fs* is in model units, kilometers, and is used by **triseis**; variable *sx* is an integer number of meters and is computed for a header value.

Variable *fg* (line M) is the geophone location relative to the start of the geophone surface. The geophone surface is defined in line Y as variable *gx*. As j increments, the geophone moves across the model. Both *fg* and *gx* are the geophone position. Variable *fg* is in model units, kilometers, and is used by **triseis**; variable *gx* is an integer number of meters and is computed for a header value.

Variable *fldr* (line G) is a header value that identifies each shot gather. Variable *tracl* (line P) sequentially numbers the traces of the line (TRACe number in the Line). Variable *tracf* (line Q) sequentially numbers the traces of each gather (TRACe number in the File).

```
A    i=0
B    while i ne 40
C    do
D
E       fs   = i * 0.05      triseis
F       sx   = i * 50        sushw
G       fldr = i + 1         sushw
H
I       j=0
J       while j ne 60
K       do
L
M         fg     = i * 0.05 + j *0.05        triseis
N         gx     = i * 50 + j * 50 - 1475    sushw
O         offset = j * 50 - 1475             sushw
P         tracl  = i * 60 + j + 1            sushw
Q         tracf  = j + 1                     sushw
R
S         echo " $sx   $gx   $fldr   $offset   $tracl   $fs   $fg"
T
U         k=2
V         while k ne 6
W         do
X
Y           triseis < $inmodel  xs=2,3.95  xg=0.525,5.425  zs=0,0 zg=0,0 \
Z                   nangle=$nangle fangle=$fangle langle=$langle \
AA                  kreflect=$k krecord=1 fpeak=12 lscale=0.5 \
BB                  ns=1 fs=$fs ng=1 fg=$fg nt=$nt dt=$dt |
CC          suaddhead nt=$nt |
DD          sushw key=dt,tracl,tracr,fldr,tracf,trid,offset,sx,gx \
EE              a=4000,$tracl,$tracl,$fldr,$tracf,1,$offset,$sx,$gx >> temp$k
FF
GG          k = k + 1
HH
II        done
JJ        j = j + 1
KK
LL      done
MM    i= i + 1
NN
OO   done
```

Variable *offset* (line O) ranges from -1475 to +1475, incrementing by 50 meters. The values of offset as it passes through zero offset are ... -125, -75, -25, +25, +75, +125, etc. This simple numbering scheme cleverly skips the geophysically impossible zero offset acquisition position.

Let's use **surange** to examine the trace headers:

```
surange  <  seis1.su
```

The **surange** output is:

```
2400 traces:
 tracl=(1,2400)  tracr=(1,2400)  fldr=(1,40)  tracf=(1,60)  trid=1
 offset=(-1475,1475)  sx=(0,1950)  gx=(-1475,3425)  ns=751 dt=4000
```

5.3   *Acquire 2-D Line from Model 2: acq2.sh*

Script *acq2.sh* acquires seismic data over model *model2.dat*, generated by script *model2.sh* (Section 4.3). The survey layout is the same as in *acq1.sh*; that is:

- Shot locations range from 2 km to 3.95 km (line 68)
- Geophone locations range from 0.525 km to 5.425 km (line 68)

For seismic data acquired by *acq2.sh*, we want header (key) *sx* to hold actual source locations and we want header *gx* to hold actual receiver locations. With this in mind, the following are lines of *acq2.sh* modified from *acq1.sh* to put actual source and receiver locations in the headers.

```
34    sx=`bc -l <<-END
35    $i * 50 + 2000
36    END`

48    gx=`bc -l <<-END
49    $i * 50 + $j * 50 + 525
50    END`
```

Let's use **surange** to examine the trace headers:

```
surange  <  seis2.su
```

The **surange** output is:

```
2400 traces:
tracl=(1,2400)  tracr=(1,2400)  fldr=(1,40)  tracf=(1,60)  trid=1
offset=(-1475,1475)  sx=(2000,3950)  gx=(525,5425)  ns=751 dt=4000
```

We see that headers *sx* and *gx* of *seis2.su* differ from *seis1.su*. All other headers are the same as before.

The only other ways *acq2.sh* differs from *acq1.sh* are lines 2 and 7:

```
2  # File: acq2.sh
7  num=2
```

That is, the file name is internally documented and Model 2 is called instead of Model 1.

5.4   *Acquire 2-D Line from Model 3: acq3.sh*

Script *acq3.sh* acquires seismic data over model *model3.dat*, generated by script *model3.sh* (Section 4.4). Source and geophone positions in *acq3.sh* are shifted one unit to the left relative to the previous acquisition scripts. We can see this by line 68:

```
68      triseis < $inmodel  xs=1,2.95  xg=-0.475,4.425  zs=0,0 zg=0,0 \
```

The survey layout is:

- Shot locations (**triseis** variable *xs*) range from 1 km to 2.95 km
- Geophone locations (**triseis** variable *xg*) range from -0.475 km to 4.425 km

For seismic data acquired by *acq3.sh*, we want header (key) *sx* to hold actual source locations and we want header *gx* to hold actual receiver locations. With this in mind, the following are lines of *acq3.sh*:

```
34    sx=`bc -l <<-END
35    $i * 50 + 1000
36    END`

48    gx=`bc -l <<-END
49    $i * 50 + $j * 50 - 475
50    END`
```

Let's use **surange** to examine the trace headers:

```
    surange  <  seis3.su
```

The **surange** output is:

```
2400 traces:
 tracl=(1,2400)  tracr=(1,2400)  fldr=(1,40)  tracf=(1,60)  trid=1
 offset=(-1475,1475)  sx=(1000,2950)  gx=(-475,4425)  ns=751 dt=4000
```

We see that headers *sx* and *gx* of *seis3.su* contain actual source and receiver locations. All other headers are the same as before.

The only other ways *acq3.sh* differs from previous acquisition scripts are line 2 and 7:

```
 2  # File: acq3.sh
 7  num=3
```

That is, the file name is internally documented and Model 3 is called.

5.5   *Computer Note 7: Pause and Resume Screen Print*

While a process, such as a script, is writing messages to the screen, and it is not running in the background (see Section 1.8.1), you can pause screen printing and resume screen printing. These actions can be done only in an active window. You can make a window active by clicking on the window frame.

1. Stop screen printing: hold down the Control key and briefly press the S key. That is, press Control-S.

2. Resume screen printing: hold down the Control key and briefly press the Q key. That is, press Control-Q.

5.6   *Examine Shot Gathers: showshot.sh*

Now that we have seismic data (output files *seis1.su*, *seis2.su*, and *seis3.su*), it is time to extract shot gathers for review. Script *showshot.sh* (below):

- windows one gather from a seismic data set (line 14),
- makes a wiggle display of the seismic data (lines 17-19),
- converts the seismic data to a .eps file (lines 22-24), then
- writes the .eps image to output (the end of line 24).

The script expects two values on the command line, one for the model, another for the shot number, as shown in lines 7 and 8.

This script assumes that all seismic data are in files of the same type of name as shown <u>underlined</u> in line 14 (*seis1.su*, *seis2.su*, etc.), and in the same directory as the script.

```
 1  #!/bin/sh
 2
 3  # showshot.sh: Window one "field record" from file seis#.su
 4  #              where # represents the model number.
 5  # Outputs: wiggle image of the shot gather
 6  #          .eps file of the shot gather
 7  # Use:     showshot.sh   model   shot
 8  # Example: showshot.sh   3       20
 9
10  # Set messages on
11  set -x
12
13  # Window one "field record" to a temporary file
14  suwind < seis$1.su key=fldr min=$2 max=$2 > temp$1$2.su
15
```

```
16   # Make wiggle plot
17   suxwigb < temp$1$2.su title="SP # $2 [$1]" key=offset \
18          label1=" Time (s)" label2="Offset" \
19          x2beg=-1500 x2end=1500 perc=99 &
20
21   # Create .eps image of a shot gather
22   supswigp < temp$1$2.su title="SP # $2 [$1]" key=offset \
23          label1="Time (s)" label2="Offset (m)" \
24          x2beg=-1500 x2end=1500 perc=99 > shot$1$2.eps &
25
26   # Remove temporary gather
27   rm -f temp$1$2.su
28
29   # Exit politely from shell
30   exit
31
```

Entering

> **showshot.sh 2 25**

commands *showshot.sh* to read file *seis2.su*, extract shot 25, and create file *shot225.eps*. That is, the output .eps file name has the number of the model and the number of the shot, shown <u>underlined</u> in line 24. The title on the output images (lines 17 and 22) contain the abbreviation SP (shot point), the number of the shot gather, and in square brackets, the number of the seismic file (Figure 5.2).

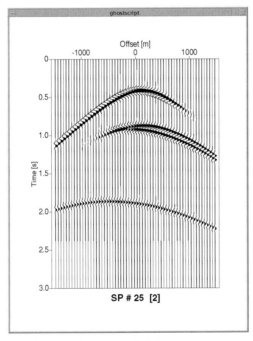

Figure 5.2: *Image of Shot Point 25 from Model 2.*

This script does not output a .su file. A temporary .su file is created on line 14, is used by **suxwigb** (line 17) and **supswigp** (line 22), and it is removed by line 27. If you want a permanent .su file:

- Change the word "temp" (lines 14, 17, and 22) to something more useful (like "shot").
- Comment out line 27 or remove lines 26 and 27.
- Add a comment line after line 5 describing the new output file.

- Remove the word "temporary" from comment line 13

Internally, the script refers to the first variable that it reads from input, model number, as *$1* and the second variable that it reads from input, shot number, is *$2*.

5.7   A Graphics Program is a Terminal Program

A graphics program (**suxwigb**, **supswigp**, etc.) is a terminal program; data cannot be piped ("|") from a graphics program to another program. In *showshot.sh*, the output of **suwind** has to go to two graphics programs, so the disk file is necessary. If a script uses only one graphics program, we can pipe the output of a processing program directly to a graphics program without making a disk file.

1. The following is acceptable:

```
process   |   graphics program
```

2a. The following is NOT acceptable:

```
process   |   graphics program 1   |   graphics program 2
```

2b. Instead, do the following:

```
process  >  disk file
graphics program 1  <  disk file
graphics program 2  <  disk file
```

5.8   Computer Note 8: Use, Modify a Command from the History List

The shell has a record of your previous commands. You can use this "history" to repeat previous commands or recall and modify previous commands.

5.8.1   Use a Command from the History List

5.8.1.1   Up Arrow Key

Press the up arrow key repeatedly to access previous commands. When you see the command you want to use, press "Enter" or "Return" to execute the command.

5.8.1.2   History List

Enter:

```
history
```

A numbered list of previous commands scrolls on the screen. To use a previous command from the history list, enter the exclamation mark (!) and the number of the command, without spaces. For example:

```
!8
```

(sometimes pronounced "bang eight" or "shriek eight") executes command 8 from the list.

5.8.2 Modify a Command from the History List

Instead of typing a long command to make a minor modification, you can recall a previous command and modify it using a combination of the arrow keys, the delete key, and the I key. Suppose your previous command was (see Section 5.6):

```
showshot.sh  2  25
```

which made a .eps file of Model 2, shot 25. Now suppose you want to make a .eps file of Model 3, shot 25. In other words, you want to change "2" to "3."

1. Press the up arrow key once to return to the previous command.
2. Press .the left arrow key several times, until the prompt is between the "2" and the "25."
3. Press the delete key once. The "2" is now erased.
4. Press the I key once. This tells the shell you want to "insert" on the line.
5. Press "3." The model number is now changed.
6. Press "Enter" or "Return" to execute the command.

If you want to make other changes on the line, like changing the shot number, use the left or right arrow keys to move the cursor to the right of the delete/insert point. Then, use "delete" or the "I" key to make your change.

Also note:

7. Pressing the "Escape" key cancels the "insert" mode.
8. If you are not in "insert" mode, you can press "U" to undo a previous change.

When you are finished changing the line, execute your new command.

5.9 Survey Quality Control

We can run an acquisition script quickly to learn if the computed header values are the ones we want. We do this by writing the values computed by the shell to a file and commenting out the SU programs. Since the SU programs are not used, the script executes quickly. A script that normally takes half an hour to run takes only four minutes when modified as described.

The scripts in this chapter were tested with the following modifications.

At the end of line 62, the redirect with append (">>") is used to write the computed values to ASCII file *survey2.txt*. If redirect had been used without append (a single ">"), only one line would be in the file because each time the script loops, new values would overwrite old values. Append puts each output line below the previous line of output.

```
61      echo " Sx=$sx  Gx=$gx    fldr=$fldr   Offset=$offset    tracl=$tracl\
62      fs=$fs    fg=$fg" >> survey2.txt

68  ##    triseis < $inmodel  xs=2,3.95  xg=0.525,5.425  zs=0,0 zg=0,0 \
69  ##         nangle=$nangle fangle=$fangle langle=$langle \
70  ##         kreflect=$k krecord=1 fpeak=12 lscale=0.5 \
71  ##         ns=1 fs=$fs ng=1 fg=$fg nt=$nt dt=$dt |
72  ##    suaddhead nt=$nt |
73  ##    sushw key=dt,tracl,tracr,fldr,tracf,trid,offset,sx,gx \
74  ##         a=4000,$tracl,$tracl,$fldr,$tracf,1,$offset,$sx,$gx >> temp$k

91  ##echo " --Sum files."
92  ##susum temp2 temp3 > tempa
93  ##susum tempa temp4 > tempb
94  ##susum tempb temp5 > $outseis
```

Since no traces are created, the script would crash if it tried to execute lines 92-94; so, we comment them out. Line 91 is commented out because we do not want to be told that summation occurred when we know it did not.

Notice that lines 68-74 and lines 91-94 are commented out with two "#." We know only one is necessary. We use two so we see where we intend comments to be temporary.

This script would have created 2400 traces. The output file contains 2400 lines and occupies less than 2 Mbytes. Below are the first ten lines and the last ten lines of *survey2.txt*.

```
Sx=2000    Gx=525     fldr=1     Offset=-1475    tracl=1       fs=0      fg=0
Sx=2000    Gx=575     fldr=1     Offset=-1425    tracl=2       fs=0      fg=.05
Sx=2000    Gx=625     fldr=1     Offset=-1375    tracl=3       fs=0      fg=.10
Sx=2000    Gx=675     fldr=1     Offset=-1325    tracl=4       fs=0      fg=.15
Sx=2000    Gx=725     fldr=1     Offset=-1275    tracl=5       fs=0      fg=.20
Sx=2000    Gx=775     fldr=1     Offset=-1225    tracl=6       fs=0      fg=.25
Sx=2000    Gx=825     fldr=1     Offset=-1175    tracl=7       fs=0      fg=.30
Sx=2000    Gx=875     fldr=1     Offset=-1125    tracl=8       fs=0      fg=.35
Sx=2000    Gx=925     fldr=1     Offset=-1075    tracl=9       fs=0      fg=.40
Sx=2000    Gx=975     fldr=1     Offset=-1025    tracl=10      fs=0      fg=.45


Sx=3950    Gx=4975    fldr=40    Offset=1025     tracl=2391    fs=1.95   fg=4.45
Sx=3950    Gx=5025    fldr=40    Offset=1075     tracl=2392    fs=1.95   fg=4.50
Sx=3950    Gx=5075    fldr=40    Offset=1125     tracl=2393    fs=1.95   fg=4.55
Sx=3950    Gx=5125    fldr=40    Offset=1175     tracl=2394    fs=1.95   fg=4.60
Sx=3950    Gx=5175    fldr=40    Offset=1225     tracl=2395    fs=1.95   fg=4.65
Sx=3950    Gx=5225    fldr=40    Offset=1275     tracl=2396    fs=1.95   fg=4.70
Sx=3950    Gx=5275    fldr=40    Offset=1325     tracl=2397    fs=1.95   fg=4.75
Sx=3950    Gx=5325    fldr=40    Offset=1375     tracl=2398    fs=1.95   fg=4.80
Sx=3950    Gx=5375    fldr=40    Offset=1425     tracl=2399    fs=1.95   fg=4.85
Sx=3950    Gx=5425    fldr=40    Offset=1475     tracl=2400    fs=1.95   fg=4.90
```

Comparing the output to line 68 (**triseis**), notice that:

- the first Sx value matches the first xs value of line 68 (2),
- the last Sx value matches the last xs value of line 68 (3.95),
- the first Gx value matches the first xg value of line 68 (0.525), and
- the last Gx value matches the last xg value of line 68 (5.425).

6. Model 4: Build, Acquire a Line, Display Gathers, QC

6.1 Introduction

In the previous two chapters, we developed three simple models and used them to acquire 2-D lines of seismic data. In this chapter, we combine some of the attributes of those models to create a fourth model and use the model to acquire a 2-D line of seismic data

In the following two sections, we explain the model and acquisition scripts in detail; you do not have to be familiar with the related scripts that are in the previous two chapters.

The next three sections of this chapter explain how to build the model, acquire seismic data, and view selected gathers. While those sections are important, the last two sections about quality control (QC) are equally important. The QC sections explain how you can acquire survey information and seismic data from selected portions of the model. You can examine preliminary survey information and seismic data to increase your confidence in your model and your acquisition script before spending time acquiring the full seismic data set.

6.2 Model 4: Earth Model – Diffractor and Pinchout

As with previous models, Model 4 layers are homogeneous and isotropic. Each layer has a single acoustic (P-wave) velocity. Let's examine script *model4.sh*. The numbers on the left are added for discussion; they are not part of the script.

```
1    #! /bin/sh
2    # File: model4.sh
3
4    # Set messages on
5    set -x
6
7    # Experiment Number
8    num=4
9
10   # Name output binary model file
11   modfile=model${num}.dat
12
13   # Name output encapsulated Postscript image file
14   psfile=model${num}.eps
15
16   # Remove previous .eps file
17   rm -f $psfile
18
19   trimodel xmin=-2 xmax=12.0 zmin=0 zmax=2.0 \
20   1 xedge=-2,12 \
21     zedge=0,0 \
22     sedge=0,0 \
23   2 xedge=-2.,0.00,2.0,4.0,6.0,8.00,10.0,12.0 \
24     zedge=0.3,0.32,0.3,0.6,0.2,0.25,0.25,0.25 \
25     sedge=0,0,0,0,0,0,0,0 \
26   3 xedge=-2.,0.0,2.0,4.0,6.0,7.0,7.5,8.00,8.50,9.0,9.50,10.,10.5,11.0,11.5,12. \
27     zedge=0.8,0.8,1.0,1.3,0.5,0.5,0.7,0.9,1.02,1.15,1.25,1.25,1.2,1.15,1.08,1.0 \
28     sedge=0,0,0,0,0,0,0,0,0,0,0,0,0,0,0,0 \
29   4 xedge=-2.,0.0,2.0,4.0,6.0,7.00,8.0,9.0,10.0,11.0,12. \
30     zedge=1.5,1.5,1.6,1.9,1.3,1.45,1.6,1.7,1.75,1.65,1.5 \
31     sedge=0,0,0,0,0,0,0,0,0,0,0 \
32   5 xedge=1.80,2.00,2.20 \
33     zedge=0.98,1.00,1.03 \
34     sedge=0,0,0 \
```

```
35   6 xedge=1.80,2.00,2.20 \
36     zedge=0.98,1.05,1.03 \
37     sedge=0,0,0 \
38   7 xedge=7.0,7.5,8.0,8.5,9.0,9.50,10.0,11.0,12. \
39     zedge=0.5,0.6,0.7,0.8,0.9,0.91,0.88,0.72,0.5 \
40     sedge=0,0,0,0,0,0,0,0,0 \
41   8 xedge=7.0,7.5,8.0,8.50,9.00,9.50,10.0,10.5,11.0,11.5,12.0 \
42     zedge=0.5,0.7,0.9,1.02,1.15,1.25,1.25,1.20,1.15,1.08,1.00 \
43     sedge=0,0,0,0,0,0,0,0,0,0,0 \
44   9 xedge=-2,12 \
45     zedge=2,2 \
46     sedge=0,0 \
47    kedge=1,2,3,4,5,6,7,8,9 \
48    sfill=0.0,0.20,0,0,0.44,0,0 \
49    sfill=0.0,0.50,0,0,0.16,0,0 \
50    sfill=0.0,1.20,0,0,0.08,0,0 \
51    sfill=0.0,1.80,0,0,0.07,0,0 \
52    sfill=2.0,1.02,0,0,0.11,0,0 \
53    sfill=10.,1.00,0,0,0.09,0,0 > $modfile
54   ##        x,z
55
56   # Create Encapsulated PostScript (EPS) image of model
57   spsplot < $modfile > $psfile \
58          gedge=0.5 gtri=2.0 gmin=0.0 gmax=5.0 \
59          title="Earth Model $num" \
60          labelz="Depth (km)" labelx="Distance (km)" \
61          wbox=5.25 hbox=0.75 dxnum=2.0 dznum=1.0
62
63   # Exit politely from shell
64   exit
65
```

This script can be divided into sets:

- System: Line 1 invokes the shell, line 5 turns on messages, and line 64 exits the shell.

- Variables: Line 8 lets us vary the number of each run as we perfect the script (2a, 2b, etc.). This number becomes part of output file names. Line 11 assigns a name to the output binary model file that is used on line 53. **The output binary model file is the prime reason for this script**. Line 14 assigns a name to the output .eps image file that is used on line 57.

- Bookkeeping: Line 17 removes a previous image file. This line is optional. On some systems, the program crashes if a .eps file with the same name already exists. Usually files are overwritten.

- Program **trimodel**: Lines 19-53 create the model. Notice that line 54 is a comment. We use line 54 to remind us that the first two entries of line *sfill* are x-z values. The very important *sfill* line is discussed below.

- Program **spsplot**: Lines 57-61 create the .eps image file.

Let's examine program **trimodel**. Program **trimodel** fills the model with triangles of $(1/velocity)^2$. While (1/velocity) is called "slowness," $(1/velocity)^2$ is called "sloth." A sloth is a slow-moving tree-dwelling mammal found in Central and South America. Program **trimodel**'s use here can be divided into five parts:

1. Line 19 defines the model dimensions.

2. The nine sets of *xedge*, *zedge*, and *sedge* triplets define layer boundaries (interfaces). Each *xedge*, *zedge*, *sedge* triplet must have the same number of values. For example, triplet 1 that defines the top of the model has two values to define the flat layer:

```
20   1 xedge=-2,12 \
21     zedge=0,0 \
22     sedge=0,0 \
```

And, parameters *xedge*, *zedge*, *sedge* for interface two have eight values (as many values as we thought necessary to define the interface):

```
23   2 xedge=-2.,0.00,2.0,4.0,6.0,8.00,10.0,12.0 \
24     zedge=0.3,0.32,0.3,0.6,0.2,0.25,0.25,0.25 \
25     sedge=0,0,0,0,0,0,0,0 \
```

Line *sedge* has only zeros because all layers are isotropic and homogeneous. Line *sedge* would have non-zero values if we want a layer to have velocity gradients.

3. Use *kedge*, line 47, to list your interface numbers. This is important for your acquisition script (the next section). An interface NOT listed here will NOT be seen by a later acquisition script.

4. In this model that has nine boundaries, there are six layers (Figure 6.1): four somewhat horizontal layers that extend across the model, a buried channel (between interfaces 5 and 6) and a pinchout (between interfaces 7 and 8)

    Therefore, there are six *sfill* lines, lines 48-53. These lines are written for isotropic, homogenous layers, which is why most of the values are zero.

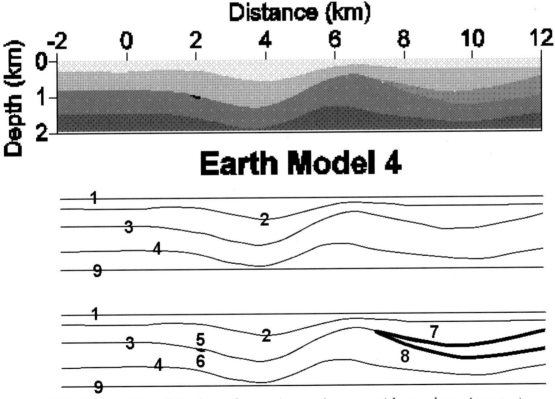

Figure 6.1: *Model 4 is a 2-D slice of a marine environment (the top layer is water). (TOP) The earth model. (MIDDLE) The simple model comprised of four layers; that is, interfaces 1 (top), 2, 3, 4, and 9 (bottom). (BOTTOM) The model with the small channel diffractor (interfaces 5 and 6) and the pinchout (interfaces 7 and 8). Interface 7 exactly matches interface 3 from x = 7 km to x = 12 km. Interfaces 7 and 8 close (pinch out) at x = 7 km.*

Table 6.1: Line *sfill* variables

x	z	x0	z0	s00	ds/dx	ds/dz

Each x-z pair of *sfill* specifies a point (any point) within a layer. Each *sfill* line defines the sloth value that fills the layer. Remember, this model contains only isotropic, homogenous layers.

Table 6.2: Values of *sfill* in Model 4

x (m)	z (m)	s00 (sloth)	velocity (1/√sloth) (m/s)	Layer boundaries
0.0	0.20	0.44	1508	1-2
0.0	0.50	0.16	2500	2-3
0.0	1.20	0.08	3536	3-4
0.0	1.80	0.07	3780	4-9
2.0	1.02	0.11	3015	5-6
10.	1.00	0.09	3333	7-8

Figure 6.2 shows that the sloth values generally decrease (P-wave velocity layer values generally increase) with depth.

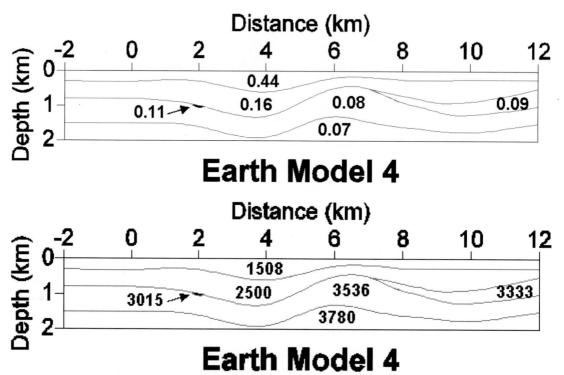

Figure 6.2: *Model 4 layer velocities. Top: Sloth values. Bottom: Velocities in m/s.*

To make interfaces 5 and 6 (below) create a layer, the two ends of the interfaces intersect (see the **bold** x,z values below).

```
32   5 xedge=1.80,2.00,2.20 \
33     zedge=0.98,1.00,1.03 \
```

```
34     sedge=0,0,0 \
35   6 xedge=1.80,2.00,2.20 \
36     zedge=0.98,1.05,1.03 \
37     sedge=0,0,0 \
```

Interfaces 7 and 8 (below) have the same starting *xedge* (7.0) and *zedge* (0.5) values (in **bold** below), ensuring that they form a pinchout. The interfaces form a closed layer because the right ends intersect the right edge of the model.

```
38   7 xedge=7.0,7.5,8.0,8.5,9.0,9.50,10.0,11.0,12. \
39     zedge=0.5,0.6,0.7,0.8,0.9,0.91,0.88,0.72,0.5 \
40     sedge=0,0,0,0,0,0,0,0,0 \
41   8 xedge=7.0,7.5,8.0,8.50,9.00,9.50,10.0,10.5,11.0,11.5,12.0 \
42     zedge=0.5,0.7,0.9,1.02,1.15,1.25,1.25,1.20,1.15,1.08,1.00 \
43     sedge=0,0,0,0,0,0,0,0,0,0,0 \
```

5. Line 53 has the variable name of the output binary model file.

Program **spsplot** plots a triangulated sloth function as an encapsulated Postscript file. So, we see that **spsplot** is designed to accompany **trimodel**! If you want to see the triangles that are used to build the model, change *gtri=2.0* to *gtri=1.0*.

On line 61, *wbox=5.25* and *hbox=0.75* ensure that the .eps image has the same proportions as the actual model (line 19). It is important for us to specify the same width and height in different programs within the same script so images created by different programs exactly overlap. Also on line 61, *dxnum* and *dznum* specify the annotation increments in the respective directions.

6.3   *Acquire 2-D Line from Model 4: acq4.sh*

In this section, we describe script *acq4.sh* that uses *model4.dat* to acquire a 2-D line of seismic data. The survey layout:

1. 200 shots (line 33)
2. Shots are equally spaced at 50m intervals (line 37 and line 40)
3. 60 split-spread traces are recorded from each shot location (line 47)
4. Geophone spacing is 50m (line 51 and line 54)
5. Geophone offsets range from -1475 m to +1475 m (line 57)
6. Shot locations range from 0 km to 9.95 km (line 75)
7. Geophone locations range from -1.475 km to 11.425 km (line 75)

The final data set has 12000 traces (200 shots x 60 geophones per shot). The generation of this data set took about thirty-six hours on a Sun UltraSPARC III with four processors. For this reason, a copy of the output file is included on the CD that accompanies this Primer.

```
 1  #! /bin/sh
 2  # File: acq4.sh
 3  # Set messages on
 4  ##set -x
 5
 6  # Assign values to variables
 7  nangle=201 fangle=-65 langle=65 nt=501 dt=0.004
 8
 9  # Model
10  num=4
11  echo " --Model number = $num"
12
13  # Name input model file
14  inmodel=model$num.dat
```

```
15
16    # Name output seismic file
17    outseis=seis${num}.su
18
19    # Remove survey file
20    rm -f survey${num}.txt
21    # Name survey file
22    survey=survey${num}.txt
23
24    #==================================================
25    # Create the seismic traces with "triseis"
26    #    i-loop = 200 source positions
27    #    j-loop = 60 geophone positions (split-spread)
28    #             per shot position
29    #    k-loop = layers 2 through 8
30    #             (do not shoot layers 1 and 9)
31
32    echo " --Begin looping over triseis."
33
34    i=0
35    while [ "$i" -ne "200" ]
36    do
37
38      fs=`bc -l <<-END
39      $i * 0.05
40      END`
41      sx=`bc -l <<-END
42      $i * 50
43      END`
44      fldr=`bc -l <<-END
45      $i + 1
46      END`
47
48      j=0
49      while [ "$j" -ne "60" ]
50      do
51
52        fg=`bc -l <<-END
53        $i * 0.05 + $j *0.05
54        END`
55        gx=`bc -l <<-END
56        $i * 50 + $j * 50 - 1475
57        END`
58        offset=`bc -l <<-END
59        $j * 50 - 1475
60        END`
61        tracl=`bc -l <<-END
62        $i * 60 + $j + 1
63        END`
64        tracf=`bc -l <<-END
65        $j + 1
66        END`
67
68        echo " Sx=$sx   Gx=$gx   fldr=$fldr   Offset=$offset   tracl=$tracl\
69        fs=$fs    fg=$fg"
70        echo " Sx=$sx   Gx=$gx   fldr=$fldr   Offset=$offset   tracl=$tracl\
71        fs=$fs    fg=$fg" >> $survey
72
73        k=2
74        while [ "$k" -ne "9" ]
75        do
76
77          triseis < $inmodel  xs=0,9.95 xg=-1.475,11.425  zs=0,0 zg=0,0 \
```

```
78                 nangle=$nangle fangle=$fangle langle=$langle \
79                 kreflect=$k krecord=1 fpeak=40 lscale=0.5 \
80                 ns=1 fs=$fs ng=1 fg=$fg nt=$nt dt=$dt |
81           suaddhead nt=$nt |
82           sushw key=dt,tracl,tracr,fldr,tracf,trid,offset,sx,gx \
83                 a=4000,$tracl,$tracl,$fldr,$tracf,1,$offset,$sx,$gx >> temp$k
84
85           k=`expr $k + 1`
86
87        done
88        j=`expr $j + 1`
89
90     done
91     i=`expr $i + 1`
92
93  done
94
95  echo " --End looping over triseis."
96
97  #==================================================
98
99  # Sum contents of the temp files
100 echo " --Sum files."
101 susum temp2 temp3 > tempa
102 susum tempa temp4 > tempb
103 susum tempb temp5 > tempc
104 susum tempc temp6 > tempd
105 susum tempd temp7 > tempe
106 susum tempe temp8 > $outseis
107
108 # Remove temp files
109 echo " --Remove temp files."
110 rm -f temp*
111
112 # Report output file
113 echo " --Output file    ** $outseis **"
114
115 # Exit politely from shell script
116 echo " --Finished!"
117 exit
118
```

This script writes messages to the screen to help us track progress:

```
11  echo " --Model number = $num"
32  echo " --Begin looping over triseis."
68     echo " Sx=$sx   Gx=$gx   fldr=$fldr   Offset=$offset   tracl=$tracl\
69     fs=$fs   fg=$fg"
95  echo " --End looping over triseis."
100 echo " --Sum files."
109 echo " --Remove temp files."
113 echo " --Output file    ** $outseis **"
116 echo " --Finished!"
```

Lines 68 and 69 write variable values to the screen. Notice that line 4 is commented out. We find that once a script is perfected, the messages from line 4 interfere with reading the values printed to the screen by lines 68 and 69.

Line 10 assigns values to variables *nangle* (the number of rays or angles that emanate from the source), *fangle* (the first angle of the fan of rays that emanate from the source), and *langle* (the last angle of the fan of rays that emanate from the source). The larger the value of *nangle*, the longer it takes to produce the image. However, a complicated model might require many rays so some will reflect to surface receivers.

Line 10 also assigns the number of time samples (*nt=501*) and the time sample interval in seconds (*dt=0.004*). The output traces are 2 seconds long.

Note: Check the unit of the time sample interval in each SU program. The unit varies.

Lines 70 and 71 are almost the same as the previous two lines. However, where lines 68 and 69 write to the screen, lines 70 and 71 write values to a disk file. Although the survey file for model 4 contains 12000 lines, it only occupies 1 Megabyte. Line 20 removes this file at the beginning of each run. If this line was not used, successive runs would be appended (>>) to the end of previous runs. The survey file is named on line 22.

```
20   rm -f survey${num}.txt
22   survey=survey${num}.txt
70       echo " Sx=$sx   Gx=$gx    fldr=$fldr    Offset=$offset    tracl=$tracl\
71       fs=$fs   fg=$fg" >> $survey
```

The core of the script is the three `do-while` loops (Section 2.3.13) over **triseis**, **suaddhead**, and **sushw**. Each time these programs are used, **triseis** creates a seismic trace, **suaddhead** creates a trace header on the trace, and **sushw** writes values to the trace headers.

- The outermost i-loop goes over 200 source positions, 0 to 199 (lines 34 and 35).
- The j-loop goes over 60 geophone positions, 0 to 59 (lines 48 and 49).
- The innermost k-loop goes over the reflectors, 2 to 8 (lines 73 and 74).

Consider the innermost loop. The value of k signifies the reflector from which **triseis** records a reflection (*kreflect=$k*, line 79) and the *temp* file in which the trace is stored (*temp$k*, line 83.) Because k loops over reflectors 2 to 8, seven *temp* files are created. (The top and base of the model are ignored.) Each *temp* data set has (in this case) 12000 traces. After all traces are created, lines 101 to 106 "sum" the seven data sets, trace-for-trace.

The following pseudo-code of the `do-while` loops is presented to help us consider how the variables are used. The program in which the variable is used is printed in bold to the right on the assignment line. Variable k is <u>underlined</u> in lines T, Y, and CC to remind us of its use as discussed in the preceding paragraph.

Variable *fs* (line E) is the source location relative to the start of the source surface. The source surface is defined in line Y as variable *xs*. As i increments, the source moves across the model. Both *fs* and *sx* are the source position. Variable *fs* is in model units, kilometers, and is used by **triseis**; variable *sx* is an integer number of meters and is computed for a header value.

Variable *fg* (line M) is the geophone location relative to the start of the geophone surface. The geophone surface is defined in line Y as variable *gx*. As j increments, the geophone moves across the model. Both *fg* and *gx* are the geophone position. Variable *fg* is in model units, kilometers, and is used by **triseis**; variable *gx* is an integer number of meters and is computed for a header value.

Variable *fldr* (line G) is a header value that identifies each shot gather. Variable *tracl* (line P) numbers the traces continuously throughout the line. Variable *tracf* (line Q) numbers the traces within each gather.

```
A   i=0
B   while i ne 200
C   do
```

```
D
E      fs   = i * 0.05       triseis
F      sx   = i * 50         sushw
G      fldr = i + 1          sushw
H
I      j=0
J      while j ne 60
K      do
L
M         fg     = i * 0.05 + j *0.05        triseis
N         gx     = i * 50 + j * 50 - 1475    sushw
O         offset = j * 50 - 1475             sushw
P         tracl  = i * 60 + j + 1            sushw
Q         tracf  = j + 1                     sushw
R
S         k=2
T         while k ne 9
U         do
V
W            triseis < $inmodel  xs=0,9.95  xg=-1.475,11.425   zs=0,0 zg=0,0 \
X                   nangle=$nangle fangle=$fangle langle=$langle \
Y                   kreflect=$k krecord=1 fpeak=40 lscale=0.5 \
Z                   ns=1 fs=$fs ng=1 fg=$fg nt=$nt dt=$dt |
AA           suaddhead nt=$nt |
BB           sushw key=dt,tracl,tracr,fldr,tracf,trid,offset,sx,gx \
CC                 a=4000,$tracl,$tracl,$fldr,$tracf,1,$offset,$sx,$gx >> temp$k
DD
EE           k = k + 1
FF
GG         done
HH         j = j + 1
II
JJ      done
KK      i= i + 1
LL
MM   done
```

Variable *offset* (line O) ranges from -1475 to +1475, incrementing by 50 meters. The values of offset as it passes through zero offset are … -125, -75, -25, +25, +75, +125, etc. This simple numbering scheme cleverly skips the geophysically impossible zero offset acquisition position.

Let's use **surange** to examine the trace headers:

```
surange  <  seis4.su
```

The **surange** output is:

```
12000 traces:
 tracl=(1,12000)  tracr=(1,12000)  fldr=(1,200)  tracf=(1,60)  trid=1
 offset=(-1475,1475)  sx=(0,9950)  gx=(-1475,11425)  ns=501 dt=4000
```

6.4   *Model-Acquisition Checklist*

The model script and the acquisition script must tie together.

- In general, always list **all** your interfaces in the model script's *kedge* parameter; for example, script *model4.sh*, line 47.

- Be sure your acquisition k-loop cycles through **all** the correct interfaces; for example, script acq4.sh, lines 73-74.

    You might have wondered why, in the model script, we always put the bottom interface as the last interface (6th out of six in *acq1.sh* or 9th out of nine in *acq4.sh*). We do it for the same reason we make the first model interface the top interface. We do it because this approach simplifies designing the k-loop (script *acq4.sh*, lines 73-87). By making the model's bottom interface the last one (and the top interface the first), the k-loop (**triseis, suaddhead, sushw**) begins at the second interface and ends before the bottom interface.

On the other hand, if we put the bottom interface somewhere in the middle; say, as interface five out of nine, this is not an overwhelming problem. We would create two k-loops:

```
k=2
while [ "$k" -ne "5" ]
do
   ---
  k=`expr $k + 1`
done
```

and

```
k=6
while [ "$k" -ne "10" ]
do
   ---
  k=`expr $k + 1`
done
```

- Be sure you create and sum (**susum**) the correct number of temp files; for example, script *acq4.sh*, lines 101-106.

6.5   *Examine Shot Gathers*

Now that we have seismic data (*seis4.su*), it is time to extract shot gathers for review. Script *showshot.sh* (below):

- windows one gather from a seismic data set (line 14),
- makes a wiggle display of the seismic data (lines 17-19),
- converts the seismic data to a .eps file (lines 22-24), then
- writes the .eps image to output (the end of line 24).

The script expects two values on the command line, one for the model, another for the shot number, as shown in lines 7 and 8.

This script assumes that all seismic data are in files of the same type of name as shown <u>underlined</u> in line 14 (*seis1.su*, *seis2.su*, etc.), and in the same directory as the script.

```
 1  #! /bin/sh
 2
 3  # showshot.sh: Window one "field record" from file seis#.su
 4  #                 where # represents the model number.
 5  # Outputs: wiggle image of the shot gather
 6  #          .eps file of the shot gather
 7  # Use:     showshot.sh   model   shot
 8  # Example: showshot.sh   3        20
 9
10  # Set messages on
11  set -x
12
13  # Window one "field record" to a temporary file
14  suwind < seis$1.su key=fldr min=$2 max=$2 > temp$1$2.su
15
```

```
16   # Make wiggle plot
17   suxwigb < temp$1$2.su title="SP # $2 [$1]" key=offset \
18           label1=" Time (s)" label2="Offset (m)" \
19           x2beg=-1500 x2end=1500 perc=99 &
20
21   # Create .eps image of a shot gather
22   supswigp < temp$1$2.su title="SP # $2 [$1]" key=offset \
23           label1="Time (s)" label2="Offset (m)" \
24           x2beg=-1500 x2end=1500 perc=99 > shot$1$2.eps &
25
26   # Remove temporary gather
27   rm -f temp$1$2.su
28
29   # Exit politely from shell
30   exit
31
```

Entering

```
showshot.sh  4   45
```

commands *showshot.sh* to read file *seis4.su*, extract shot 45, and create file *shot445.eps*. That is, the output .eps file name has the number of the model and the number of the shot, as shown <u>underlined</u> in line 24. The title on the output images (lines 17 and 22) contains the abbreviation SP (shot point), the number of the shot gather, and in square brackets, the number of the seismic file (Figure 6.3).

This script does not output a .su file. A temporary .su file is created on line 14, is used by **suxwigb** (line 17) and **supswigp** (line 22), and is removed by line 27. If you want a permanent .su file:

- Change the word "temp" (lines 14, 17, and 22) to something more useful (like "shot").
- Comment out line 27 or remove lines 26 and 27.
- Add a comment line above line 5 describing the new output file.
- Remove the word "temporary" from comment line 13

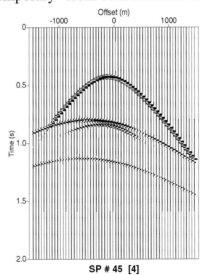

Figure 6.3: Shot gather 45 from Model 4, shot almost directly over the channel.

The images below show selected shot gathers acquired over Model 4.

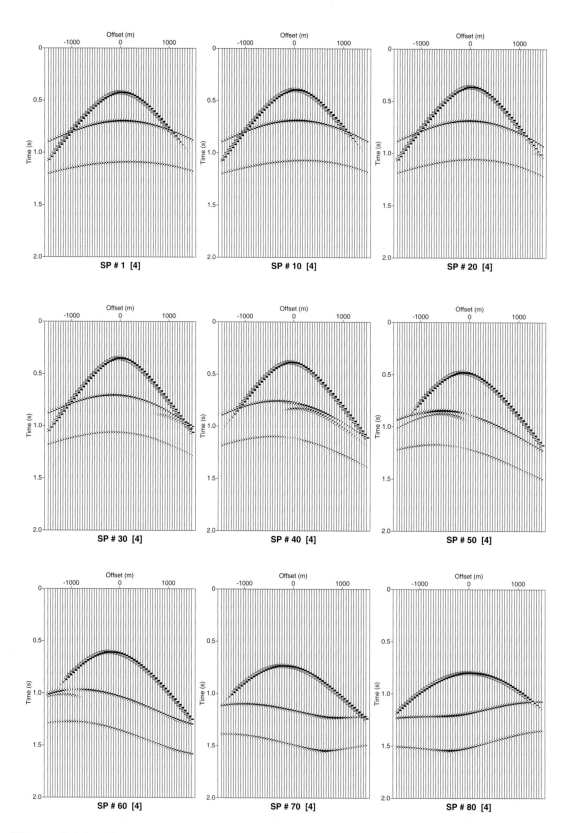

Figure 6.4.1: *Shot gathers from Model 4. Numbers correspond to shot points.*

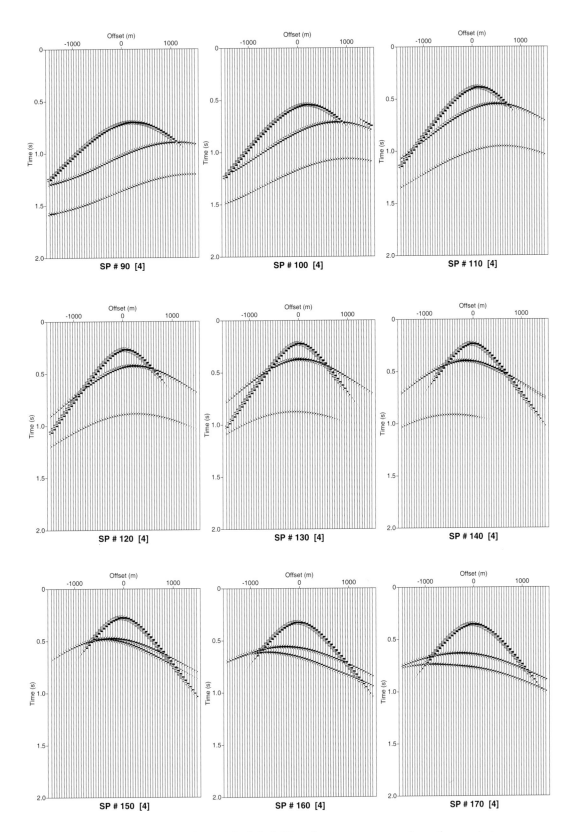

Figure 6.4.2: *Shot gathers from Model 4. Numbers correspond to shot points.*

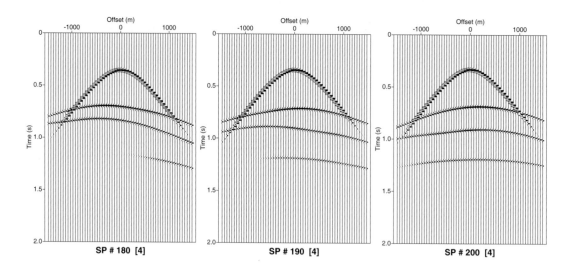

Figure 6.4.3: *Shot gathers from Model 4. The channel creates the diffraction in SP 40 and SP 50. We first see the right side pinch-out in SP 150. Due to divergent ray paths, the bottom layer does not return reflections in SP 150, SP 160, and SP 170.*

6.6 Computer Note 9: Data Files Not in Script Directory

As stated in the previous section, script, *showshot.sh*, is written with the assumption that our data files are in the same directory as the script. Let's simplify and revise the script to illustrate reading and writing data files that are not in the same directory as the script. The scripts below read the seismic data output by the acquisition script, write a selected shot gather to disk, and make a wiggle display of the single gather (**r**ead-**w**rite-**d**isplay).

Let's start by learning two pointers:

- .   a single period refers to the current directory.
- ..   two periods together refers to the directory directly above the current directory.

If you enter:

```
ls   -a
```

among other files listed you will see these two pointers. Every Unix directory is connected to the others by these pointers. You can think of the single period as "here" and the double period as "up one directory." We can use these pointers instead of using full path names in scripts (and line commands).

6.6.1 Data Directory below Script Directory — Full Path Name

We sometimes find it convenient to develop our scripts in a directory separate from our data. In this case, it is easy to refer to a data directory directly below the current directory. Before showing the easy way, we show the full path name way.

```
1   #! /bin/sh
2
3   # shotrwd1.sh: Window one "field record" from file seis#.su
4   #              where # represents the model number.
5   # Outputs: .su file of the shot gather
6   #          wiggle image of the shot gather
7   # Use:     shotrwd1.sh   model   shot
8   # Example: shotrwd1.sh   3       20
```

```
 9
10   # Set messages on
11   set -x
12
13   # Window one "field record"
14   suwind < /home/forel/suscripts/data/seis$1.su \
15         key=fldr min=$2 max=$2 \
16         > /home/forel/suscripts/data/shot$1$2.su
17
18   # Make wiggle plot
19   suxwigb < /home/forel/suscripts/data/shot$1$2.su \
20         title="SP # $2 [$1]" key=offset \
21         label1=" Time (s)" label2="Offset (m)" \
22         x2beg=-1500 x2end=1500 perc=99 &
23
24   # Exit politely from shell
25   exit
26
```

Entering

shotrwd1.sh   4   130

commands *shotrwd1.sh* to read file *seis4.su* that is in subdirectory *data* (line 14), extract shot 130, write file *shot4130.su* to subdirectory *data* (line 16), read the single gather from directory *data* (line 19), and display field record 130 as a wiggle plot.

6.6.2   Data Directory below Script Directory — Short Path Name

Assume we are in directory */home/forel/suscripts*. Let's rewrite the script using short path names. Remember to change the script's file name and the internal references to this file name (lines 3, 7, and 8).

```
 1   #! /bin/sh
 2
 3   # shotrwd2.sh: Window one "field record" from file seis#.su
 4   #              where # represents the model number.
 5   # Outputs: .su file of the shot gather
 6   #          wiggle image of the shot gather
 7   # Use:     shotrwd2.sh  model  shot
 8   # Example: shotrwd2.sh  3       20
 9
10   # Set messages on
11   set -x
12
13   # Window one "field record"
14   suwind < ./data/seis$1.su \
15         key=fldr min=$2 max=$2 \
16         > ./data/shot$1$2.su
17
18   # Make wiggle plot
19   suxwigb < ./data/shot$1$2.su \
20         title="SP # $2 [$1]" key=offset \
21         label1=" Time (s)" label2="Offset (m)" \
22         x2beg=-1500 x2end=1500 perc=99 &
23
24   # Exit politely from shell
25   exit
26
```

Entering

shotrwd2.sh   4   150

commands *shotrwd2.sh* to start here (*./*), go to subdirectory *data*, read file *seis4.su* (line 14), extract shot 130, write file *shot4150.su* to subdirectory *data* that is below the current directory (*./*) (line 16). Then (line 19) says, from here (*./*), go to subdirectory *data*, read the single gather of field record 150 and display it as a wiggle plot.

Note: You do not need to use *./* in the above script. The following script is just as effective.

```
 1   #! /bin/sh
 2
 3   # shotrwd3.sh: Window one "field record" from file seis#.su
 4   #                where # represents the model number.
 5   # Outputs: .su file of the shot gather
 6   #           wiggle image of the shot gather
 7   # Use:      shotrwd3.sh   model   shot
 8   # Example: shotrwd3.sh   3       20
 9
10   # Set messages on
11   set -x
12
13   # Window one "field record"
14   suwind < data/seis$1.su \
15          key=fldr min=$2 max=$2 \
16          > data/shot$1$2.su
17
18   # Make wiggle plot
19   suxwigb < data/shot$1$2.su \
20          title="SP # $2 [$1]" key=offset \
21          label1=" Time (s)" label2="Offset (m)" \
22          x2beg=-1500 x2end=1500 perc=99 &
23
24   # Exit politely from shell
25   exit
26
```

6.6.3 Data Directory at the same Level as Script Directory

Suppose you have a directory *worksu* and under *worksu* are two directories: *scripts* and *data*. Our problem is to make a script that is in directory "*scripts*" read the data directory without using full path names. Here is the solution:

```
 1   #! /bin/sh
 2
 3   # shotrwd4.sh: Window one "field record" from file seis#.su
 4   #                where # represents the model number.
 5   # Outputs: .su file of the shot gather
 6   #           wiggle image of the shot gather
 7   # Use:      shotrwd4.sh   model   shot
 8   # Example: shotrwd4.sh   3       20
 9
10   # Set messages on
11   set -x
12
13   # Window one "field record"
14   suwind < ../data/seis$1.su \
15          key=fldr min=$2 max=$2 \
16          > ../data/shot$1$2.su
17
18   # Make wiggle plot
19   suxwigb < ../data/shot$1$2.su \
20          title="SP # $2 [$1]" key=offset \
21          label1=" Time (s)" label2="Offset (m)" \
```

```
22          x2beg=-1500 x2end=1500 perc=99 &
23
24  # Exit politely from shell
25  exit
26
```

Our script is in directory *scripts*. On line 14, "*../*" directs the system to go up one directory (to directory *worksu*). The rest of the path name specifies the data file going down from directory *worksu*.

You can use "*../*" repeatedly to climb up the directory structure. For example,

```
suwind < ../../worksu/data/seis$1.su \
```

tells the system to go from directory *scripts* up to *worksu*, up to *forel*, down to *worksu*, down to *data*. Don't specify file names going up.

6.7   *Quality Control: The Survey File*

We can run an acquisition script quickly to learn if the computed header values are the ones we want. We do this by commenting out the SU programs, running the acquisition script, then examining the output survey file. Since the SU programs are not used, the script executes quickly. The acquisition script that took thirty-six hours takes only half an hour when we comment out lines 77-83.

```
77  ##    triseis < $inmodel  xs=0,9.95 xg=-1.475,11.425  zs=0,0 zg=0,0 \
78  ##          nangle=$nangle fangle=$fangle langle=$langle \
79  ##          kreflect=$k krecord=1 fpeak=40 lscale=0.5 \
80  ##           ns=1 fs=$fs ng=1 fg=$fg nt=$nt dt=$dt |
81  ##    suaddhead nt=$nt |
82  ##    sushw key=dt,tracl,tracr,fldr,tracf,trid,offset,sx,gx \
83  ##          a=4000,$tracl,$tracl,$fldr,$tracf,1,$offset,$sx,$gx >> temp$k

100 ##echo " --Sum files."
101 ##susum temp2 temp3 > tempa
102 ##susum tempa temp4 > tempb
103 ##susum tempb temp5 > tempc
104 ##susum tempc temp6 > tempd
105 ##susum tempd temp7 > tempe
106 ##susum tempe temp8 > $outseis

109 ##echo " --Remove temp files."
110 ##rm -f temp*

113 ##echo " --Output file   ** $outseis **"
```

Since no traces are created (no temp files), the script would crash if it tried to execute lines 101-106; so, we comment them out. If executed, line 110 would not make the script crash, but there is nothing wrong with commenting it out. We comment out lines 100, 109, and 113 because we do not want to read false information.

Notice that these lines are commented out with two "#." We know only one is necessary. We use two so we see where we intend comments to be temporary.

Below are the first ten lines and the last ten lines of *survey4.txt*.

```
Sx=0   Gx=-1475   fldr=1   Offset=-1475   tracl=1   fs=0   fg=0
Sx=0   Gx=-1425   fldr=1   Offset=-1425   tracl=2   fs=0   fg=.05
Sx=0   Gx=-1375   fldr=1   Offset=-1375   tracl=3   fs=0   fg=.10
Sx=0   Gx=-1325   fldr=1   Offset=-1325   tracl=4   fs=0   fg=.15
Sx=0   Gx=-1275   fldr=1   Offset=-1275   tracl=5   fs=0   fg=.20
Sx=0   Gx=-1225   fldr=1   Offset=-1225   tracl=6   fs=0   fg=.25
Sx=0   Gx=-1175   fldr=1   Offset=-1175   tracl=7   fs=0   fg=.30
```

```
Sx=0      Gx=-1125    fldr=1     Offset=-1125    tracl=8       fs=0      fg=.35
Sx=0      Gx=-1075    fldr=1     Offset=-1075    tracl=9       fs=0      fg=.40
Sx=0      Gx=-1025    fldr=1     Offset=-1025    tracl=10      fs=0      fg=.45

Sx=9950   Gx=10975    fldr=200   Offset=1025     tracl=11991   fs=9.95   fg=12.45
Sx=9950   Gx=11025    fldr=200   Offset=1075     tracl=11992   fs=9.95   fg=12.50
Sx=9950   Gx=11075    fldr=200   Offset=1125     tracl=11993   fs=9.95   fg=12.55
Sx=9950   Gx=11125    fldr=200   Offset=1175     tracl=11994   fs=9.95   fg=12.60
Sx=9950   Gx=11175    fldr=200   Offset=1225     tracl=11995   fs=9.95   fg=12.65
Sx=9950   Gx=11225    fldr=200   Offset=1275     tracl=11996   fs=9.95   fg=12.70
Sx=9950   Gx=11275    fldr=200   Offset=1325     tracl=11997   fs=9.95   fg=12.75
Sx=9950   Gx=11325    fldr=200   Offset=1375     tracl=11998   fs=9.95   fg=12.80
Sx=9950   Gx=11375    fldr=200   Offset=1425     tracl=11999   fs=9.95   fg=12.85
Sx=9950   Gx=11425    fldr=200   Offset=1475     tracl=12000   fs=9.95   fg=12.90
```

Comparing the output to line 77 (**triseis**), observe that:

- the first Sx value corresponds to the first xs value of line 77 (0),
- the last Sx value corresponds to the last xs value of line 77 (9.95),
- the first Gx value corresponds to the first xg value of line 77 (-1.475), and
- the last Gx value corresponds to the last xg value of line 77 (11.425).

6.8 Quality Control: Acquire and Display Specific Shot Gathers

Before shooting the entire line, you might want to acquire single shot gathers to preview seismic data and header values. It is surprisingly simple to modify the acquisition script to do this.

- Copy script *acq4.sh* to a new file and document that change in line 2 of the new script.

```
2   # File: acq4shot.sh
```

- Change the names of the output files. Line 17 contains the name of the output seismic file. Lines 20, 22, and 71 contain the name of the survey file.

```
17   outseis=seis${num}shot.su
20   rm -f survey${num}shot.txt
22   survey=survey${num}shot.txt
70       echo " Sx=$sx   Gx=$gx    fldr=$fldr   Offset=$offset   tracl=$tracl\
71       fs=$fs    fg=$fg" >> $survey
```

- Most important, specify the shot gather using lines 34 and 35. Here, we acquire gather 50:

```
34   i=49
35   while [ "$i" -ne "50" ]
```

No other changes are necessary.

A similar change acquires several shot gathers together. The following two lines acquire gathers 51-55.

```
34   i=50
35   while [ "$i" -ne "55" ]
```

If you do not change output file names, you run the risk of overwriting valuable data during future runs.

The script below, *showshotB.sh*, is designed to accompany script *acq4shot.sh*. Lines 3 and 14 differ from script *showshot.sh* in specifying the new input file.

```
1   #! /bin/sh
2
3   # showshotB.sh: Window one "field record" from file seis#shot.su
4   #                where # represents the model number.
5   # Outputs: wiggle image of the shot gather
```

```
 6  #            .eps file of the shot gather
 7  # Use:      showshotB.sh  model  shot
 8  # Example: showshotB.sh  3      20
 9
10  # Set messages on
11  set -x
12
13  # Window one temporary "field record"
14  suwind < seis$1shot.su key=fldr min=$2 max=$2 > temp$1$2.su
15
16  # Make wiggle plot
17  suxwigb < temp$1$2.su title="SP # $2 [$1]" key=offset \
18          label1=" Time (s)" label2="Offset (m)" \
19          x2beg=-1500 x2end=1500 perc=99 &
20
21  # Create .eps image of a shot gather
22  supswigp < temp$1$2.su title="SP # $2 [$1]" key=offset \
23          label1="Time (s)" label2="Offset (m)" \
24          x2beg=-1500 x2end=1500 perc=99 > shot$1$2.eps &
25
26  # Remove temporary gather
27  rm -f temp$1$2.su
28
29  # Exit politely from shell
30  exit
31
```

Entering

showshotB.sh 4   50

commands *showshotB.sh* to read file *seis4shot.su*, extract shot 50, and create file *shot450.eps*. That is, the output .eps file name has the number of the model and the number of the shot, as highlighted in line 24. The title on the output images (lines 17 and 22) contains the abbreviation SP (shot point), the number of the shot gather, and in square brackets, the number of the seismic file (Figure 6.3).

This script does not output a .su file. A temporary .su file is created on line 14, is used by **suxwigb** (line 17) and **supswigp** (line 22), and is removed by line 27. If you want a permanent .su file, follow the instructions in section 6.5.

7. Model 4: Sort, Velocity Analysis

7.1 Introduction

From the previous chapter, we have a 2-D line of seismic data. In this chapter, we:

- sort the shot gathers to common midpoint (CMP) gathers and
- perform velocity analysis on selected CMP gathers.

Because our synthetic data are noise-free, we can proceed quickly from shot gathers to migration. If we wanted to, we could use **suaddnoise** to add noise to the data; this is often useful. You should consider adding some noise to the data set, then processing it using the examples we provide in this chapter and the next.

Let's discuss "common-depth-point" (CDP) and "common midpoint" (CMP) (Sheriff, 2002). We try to avoid the term CDP because there is no common (same) point at the reflector if the reflector dips. On the other hand, common-midpoints almost always exist because they are defined by the geometric midpoint between sources and receivers. Because SU does not have a *cmp* key, we reluctantly use the *cdp* key.

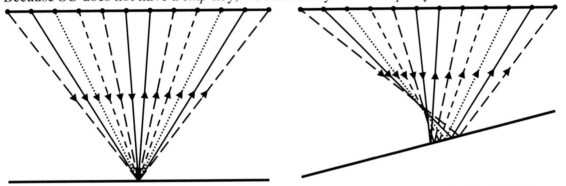

Figure 7.1: *Left: The common-depth-point (CDP) and the common-midpoint (CMP) are the same because the reflector is flat. Right: The reflections in this CMP gather do not share a CDP because the reflector dips.*

7.2 Sort from Shot Gathers to CMP Gathers

Script *sort2cmp.sh* does two jobs:

1. We use program **suchw** (Change Header Word) to create header (key) *cdp* and assign values to it. Using the equation below, values of *cdp* (key1) are computed from *gx* (key2) and *sx* (key3).

   ```
   key1 = ( a + b*key2 + c*key3 ) / d
   ```

   Scalars a, b, c, and d need to be determined by sketching the geometry. Our geometry dictates a = 1525, b =1, c = 1, and d = 50. Therefore,

   ```
   cdp = ( 1525 + 1*gx + 1*sx ) / 50
   ```

2. We use program **susort** to sort the traces based on primary sort key *cdp* and secondary sort key *offset*. A secondary sort is the order within the primary sort entity.

```
1  #! /bin/sh
2  # sort2cmp.sh
3
```

```
 4  # Compute CMP header
 5  suchw < seis4.su key1=cdp key2=gx key3=sx \
 6          a=1525 b=1 c=1 d=50 |
 7
 8  # Sort data to CMP gathers
 9  #          file    1    2
10  susort > cmp4.su cdp offset
11
12  # Exit politely from shell
13  exit 0
14
```

Let's try to understand how the first part of the script works. Typically, when we assign *cdp* values, we want the output CMP gathers to increment by "1."

Recall that receiver positions increment by 50 meters and the source positions also increment by 50 meters. In the equation above, scalar "a" is set to 50 meters greater than the largest positive offset. Scalar "d" is chosen to normalize the values computed in the numerator so *cdp* values increment simply.

Table 7.1 shows headers *sx* and *gx* of the first and last traces of the first ten shots, as well as the computed cdp values. You can see that *cdp* values slowly increase in a regular way.

Table 7.1: *cdp* values computed from headers *sx* and *gx*

		cdp	sx	gx
Shot 1	tracf = 1	1	0	-1475
	tracf = 60	60	0	1475
Shot 2	tracf = 1	3	50	-1425
	tracf = 60	62	50	1525
Shot 3	tracf = 1	5	100	-1375
	tracf = 60	64	100	1575
Shot 4	tracf = 1	7	150	-1325
	tracf = 60	66	150	1625
Shot 5	tracf = 1	9	200	-1275
	tracf = 60	68	200	1675
Shot 6	tracf = 1	11	250	-1225
	tracf = 60	70	250	1725
Shot 7	tracf = 1	13	300	-1175
	tracf = 60	72	300	1775
Shot 8	tracf = 1	15	350	-1125
	tracf = 60	74	350	1825
Shot 9	tracf = 1	17	400	-1075
	tracf = 60	76	400	1875
Shot 10	tracf = 1	19	450	-1025
	tracf = 60	78	450	1925

Table 7.2 shows headers *sx* and *gx* of the first and last traces of the last ten shots, as well as the computed *cdp* values. You can see that *cdp* values slowly increase. Also, notice that the last *cdp* value is 458 – recall that there are 176 shot gathers!

Table 7.2: *cdp* values computed from headers *sx* and *gx*

		cdp	sx	gx
Shot 191	tracf = 1	381	9500	8025
	tracf = 60	440	9500	10975
Shot 192	tracf = 1	383	9550	8075
	tracf = 60	442	9550	11025

Shot 193	tracf = 1	385	9600	8125
	tracf = 60	444	9600	11075
Shot 194	tracf = 1	387	9650	8175
	tracf = 60	446	9650	11125
Shot 195	tracf = 1	389	9700	8225
	tracf = 60	448	9700	11175
Shot 196	tracf = 1	391	9750	8275
	tracf = 60	450	9750	11225
Shot 197	tracf = 1	393	9800	8325
	tracf = 60	452	9800	11275
Shot 198	tracf = 1	395	9850	8375
	tracf = 60	454	9850	11325
Shot 199	tracf = 1	397	9900	8425
	tracf = 60	456	9900	11375
Shot 200	tracf = 1	399	9950	8475
	tracf = 60	458	9950	11425

Below is the **surange** output of file *cmp4.su*.

```
        surange  <  cmp4.su
12000 traces:
 tracl=(1,12000)  tracr=(1,12000)  fldr=(1,200)  tracf=(1,60)  cdp=(1,458)
 trid=1 offset=(-1475,1475)  sx=(0,9950)  gx=(-1475,11425)  ns=501
 dt=4000
```

Below is the output of the first ten lines of command:

```
        sugethw  <  cmp4.su  key=cdp,offset,fldr
            cdp=1    offset=-1475    fldr=1
            cdp=2    offset=-1425    fldr=1
            cdp=3    offset=-1475    fldr=2
            cdp=3    offset=-1375    fldr=1
            cdp=4    offset=-1425    fldr=2
            cdp=4    offset=-1325    fldr=1
            cdp=5    offset=-1475    fldr=3
            cdp=5    offset=-1375    fldr=2
            cdp=5    offset=-1275    fldr=1
            cdp=6    offset=-1425    fldr=3
```

This list helps us see that the data are in *cdp* order (key *fldr* is no longer the primary sort order) with secondary key of *offset*. We also see that fold slowly increases: there is one line for cdp = 1 and one line for cdp = 2; there are two lines for cdp = 3 and two lines for cdp = 4; there are three lines for cdp = 5, etc.

The output of **surange** shows that the *cdp* values range from "1" to "458" (underlined above), confirming our earlier calculations.

7.3   *Extract CMP Gathers*

Now that we have a line of CMP gathers (*cmp4.su*), let's extract individual CMPs for view and for the next step: velocity analysis. Script *showcmp.sh* (below):

- windows one gather from the line and write it to disk (line 20),
- makes a wiggle display of the .su CMP gather (lines 22-24),
- converts the .su CMP gather to a .eps file (lines 26-28), then
- writes the .eps image to disk (the end of line 28).

The script expects two values on the command line, one for the model, another for the CMP number, as shown in lines 6 and 7.

This script assumes that all seismic data are in files of the same type of name as underlined in line 20 (*cmp1.su*, *cmp2.su*, etc.), and in the same directory as the script.

```
 1   #! /bin/sh
 2
 3   # File: showcmp.sh: Window one CMP from file seis#.su
 4   #                   where # represents the model number.
 5   # Specify input model number and output CMP
 6   # Input:  CMP file
 7   # Outputs: CMP gather in .su format
 8   #          wiggle image of the CMP gather
 9   #          .eps file of the CMP gather
10   # Use:      showcmp.sh [model] [cmp]
11   # Example: showcmp.sh   4      20
12
13   # Set messages on
14   ##set -x
15
16   # Display a CMP gather between 1 and 458
17   if [ $2 -le 458 ]; then
18     if [ $2 -ge 1 ]; then
19
20       suwind < cmp$1.su key=cdp min=$2 max=$2 > cmp$1$2.su
21
22       suxwigb < cmp$1$2.su title="CMP # $2 [$1]" key=offset \
23               label1=" Time (s)" label2="Offset (m)" \
24               perc=99 &
25
26       supswigp < cmp$1$2.su title="CMP # $2 [$1]" key=offset \
27                label2="Time (s)" label2="Offset (m)" \
28                x2beg=-1500 x2end=1500 perc=99 > cmp$1$2.eps
29
30       exit
31
32     fi
33   fi
34
35   echo usage: showcmp.sh [model number] [CMP between 1 and 458]
36
37   # Exit politely from shell
38   exit
39
```

7.4 Computer Note 10: If Blocks

The script above (*showcmp.sh*) uses two nested if blocks (Section 2.3.13). Script *showshot.sh* (Section 6.5) did not use if blocks, and they are not necessary here, but we take this opportunity to illustrate them.

The if blocks ensure that we input a CMP value (variable *$2*) that is within the range of actual values. Line 15 requires that the value of *$2* is 458 or less; line 16 requires that the value of *$2* is 1 or greater.

If an invalid value is input for *$2*, for example:

showcmp.sh   4   900

the "echo" message of line 33 is written to the screen:

usage: showcmp.sh [model number] [CMP between 1 and 458]

If a valid value for *$2* is input, the "echo" message is skipped.

If you think there is a problem with the script or how you are using the script, uncomment line 14.

7.5   Fold

After sorting to CMP-order, we have 458 gathers. But, the first gathers and the last gathers of the line are not full fold. See Figure 7.2, below. (We describe the program that generated the data for Figure 7.2, **sukeycount**, in Appendix C.) Upon inspection, we find that the first full-fold CMP gather is CMP 60 and the last full fold CMP gather is CMP 400. Full fold is 30 traces, in contrast to the shot gather full fold of 60 traces. This is an expected outcome of sorting based on the acquisition geometry (see Sheriff, 2002: Stacking Chart).

To make reasonably detailed velocity analysis at a regular interval along the line, we choose to analyze eighteen full-fold CMPs: 60, 80, 100, ..., 360, 380, 400.

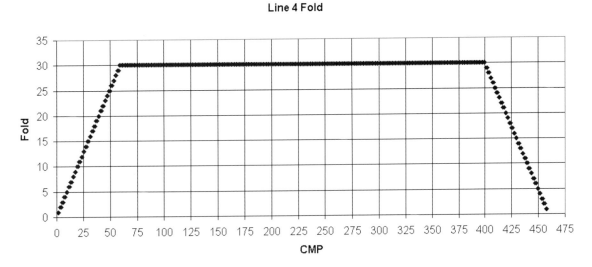

Figure 7.2: *Fold values for data set cmp4.su.*

7.6   Stacking Velocity Analysis

There are several types of velocities in seismic reflection work: interval velocity, root-mean-square (rms) velocity, stacking velocity, and migration velocity are some. At this stage of processing, we want to determine velocities that allow us to stack each CMP gather into a single trace. To reinforce primary reflections in the (following) stacking process, we want to determine a time-velocity (t-v) function that flattens reflections in each CMP that we analyze.

The following sections describe two tools that we later incorporate into an interactive velocity analysis (iva) script: the semblance plot and constant velocity stacks.

7.6.1 Prepare a Shot Gather for Velocity Analysis

To demonstrate how we prepare a real (non-synthetic) gather for semblance, we download file *ozdata.14* from the CWP ftp site. Following our file naming convention, we changed the file name to *oz14.su*: This file has 6.3 seconds of data, more than we need for our demonstration (see Figure 7.3 below).

Below is the **surange** output of file *oz14.su*.

```
        surange  <  oz14.su
48 traces:
 tracl=(1,48)  tracr=(1,48)  fldr=10014 tracf=(1,48)  cdp=(14,61)
 cdpt=1 trid=1 nvs=4 nhs=1 duse=1
 scalel=1 scalco=1 counit=1 delrt=4 muts=4
 ns=1575 dt=4000
```

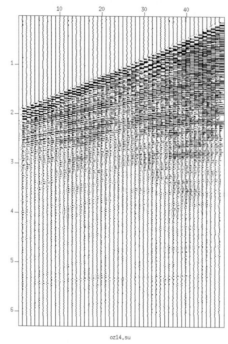

oz14.su

Figure 7.3: *Original data file oz14.su has 48 traces and is 6.3 seconds long.*

The following script, *oz14prep.sh*, prepares *oz14.su* for velocity analysis.

```
 1  #! /bin/sh
 2
 3  # File: oz14prep.sh: Prepare oz14.su for velocity analysis
 4  #        Input (1): shot gather
 5  #        Output (1): modified CMP gather
 6  #        Use: oz14prep.sh
 7
 8  # Set messages on
 9  set -x
10
11  # Name data sets
12  indata=oz14.su
13  outdata=oz14h.su
14
15  # Use only first 4 seconds of data
16  suwind < $indata tmax=4  |
17
18  # Gain
19  sugain tpow=2  |
20
21  # Convert from shot gather to CMP gather
22  #    by making all cdp values = 14
23  suchw key1=cdp a=14 b=0  |
24
25  # Add offset values to offset key
26  suchw key1=offset a=11250 b=-220 key2=tracf  |
27
28  # Mute
```

```
29   sumute key=tracf           \
30   xmute=1,46,48              \
31   tmute=2.250,0.500,0.004 \
32         > $outdata
33
34   # Plot
35   suxwigb < $outdata key=offset perc=90 &
36
37   # Exit politely from shell
38   exit
39
```

In *oz14prep.sh*, the data are prepared in the following ways:

- Line 16 windows the first 4 seconds of data.

- Line 19 applies a classic power of 2 time-varying gain.

- Line 23 "effectively" converts the shot gather to a CMP gather by making all values of key *cdp* the same. (See Section 3.7.)

- Line 26 puts values into key *offset* according to either of Yilmaz' books: Table 1-8 of Seismic Data Processing (1987) or Table 1-13 of Seismic Data Analysis (2001): the nearest offset trace is 69 feet from the source and trace spacing is 220 feet. Since the first trace is the farthest offset trace, parameter *a* has the farthest offset and parameter *b* is used to subtract the trace spacing from parameter *a*.

- Lines 29-31 mute the refractions. Figure 7.4, below left, shows the data after oz14prep.sh. For comparison, the right side of Figure 7.4 shows the data after all the processing of oz14prep.sh, but without the mute.

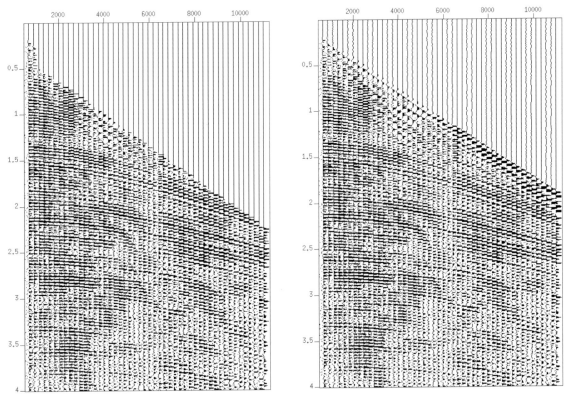

Figure 7.4: *Left: Output of script oz14prep.sh. Right: Same as Left, but without mute.*

7.6.2 Semblance Contour Plot

Velocity analysis, a necessary prelude to NMO correction, yields the stacking velocities. NMO correction is also called flattening the data. Once CMP gathers are flattened, they can be stacked. Stacking attenuates random noise and multiples. Stacking yields a zero-offset section (neglecting elastic or amplitude-versus-offset effects) that is ready to be migrated. (We ignore dip moveout correction (DMO).) After migration, the data are interpreted for geological meaning.

We would like to decrease the time it takes to do velocity analysis. One way is to limit the range of velocities that we consider to be in the data. The range of velocities is usually limited by intelligent guessing. We know that a hyperbola with a lot of curvature represents a low velocity and a hyperbola with little curvature represents a high velocity. Figure 7.5 (below) is CMP 60 from model 4. There are three obvious reflectors on this gather. We know that the upper red curve represents a velocity less than any reflector. We also know that the lower red curve represents a velocity greater than any reflector. We can use this knowledge to limit the number of velocities to compute. Also, when we know the top layer is water, as in Model 4, the slowest velocity is 1500 m/s or 4920 f/s.

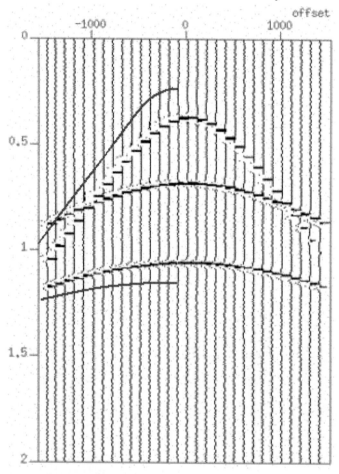

Figure 7.5: *CMP 60 from Model 4. The lower and upper lines (hyperbolas) are the lower and upper velocities necessary for velocity analysis.*

(Remember that color figures and those requiring detail are included as TIFF files on a CD accompanying this book, as listed in Appendix E.)

Unfortunately, it is not yet possible to push a button and have the computer figure out the velocities that flatten the primary reflections. Remember, real seismic data have multiples and noise that interfere with moveout analysis. The synthetic gather in Figure 7.5 also has noise, but not so much that we can't easily see the reflectors.

When we think we have a reasonable velocity range, we create a plot of a velan (VELocity ANalysis) of the CMP. SU program **suvelan** computes stacking velocities. We send (pipe) the output of **suvelan** to **suxcontour** to make a plot. Then, we pick time-velocity (t-v) pairs from the velan.

Below is script *oz14velan.sh* that creates a velan from CMP *oz14h.su*. File *oz14h.su* has noise and multiples, so velocity analysis is not easy.

```
1   #! /bin/sh
2
3   # File: oz14velan.sh: Create semblance display of one CMP
4   #        Input (1): One cdp gather
5   #        Output (1): Velan (contour) plot
6   #        Use: oz14velan.sh
7
8   # Set messages on
9   set -x
10
11  # Input data
12  indata=oz14h.su
13
14  # Important processing variables
15  # nv = number of vels, dv = delta vel, fv = first vel
16  nv=160
17  dv=50
18  fv=4000
19
20  # Do the work
21
22  suvelan < $indata nv=$nv dv=$dv fv=$fv smute=1.5 |
23  suxcontour bclip=0.5 wclip=0.0 f2=$fv d2=$dv \
24            label1=" Time (s)" label2="Velocity" \
25            title="Semblance plot for CMP $indata" \
26            windowtitle="Semblance: $indata" \
27            grid1=solid grid2=solid cmap=default \
28            nc=7 nlabelc=0 &
29
30  # Exit politely from shell
31  exit
32
```

Figure 7.6 is the velan generated by script *oz14velan.sh*. The core mathematics in **suvelan** is called semblance. The velan display is actually a plot of semblance. Semblance is defined in the **suvelan** selfdoc.

Here, we describe the velan display without mathematics. The input CMP has offset values in the trace headers. The time of each data sample is also known. Program **suvelan** goes through the times of a fictitious zero-offset trace. At each time on this zero-offset trace, it applies NMO and stack along a series of hyperbolas – the hyperbolas that correspond to velocities from 4000 to 12000 (= (50 x 160) + 4000) f/s. It does this in increments of 50 (the value of dv).

On velans, the time axis specifies zero-offset time. The display is a contour plot of amplitude reinforcement from applying moveout and stack using the range of velocities.

Here we use **suxcontour** to show black contour lines on a white background. We could also use **suximage** to create color-area displays on a computer monitor.

Parameter *nc* is a sort of volume control – by varying the number of contour levels, we coarsen or fine-tune the output of the semblance calculation.

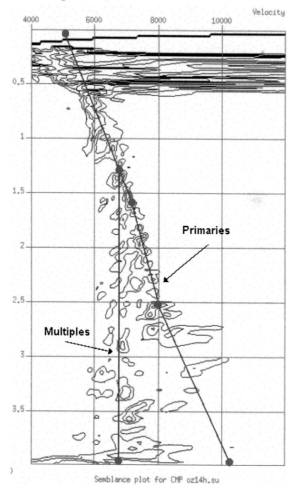

Figure 7.6: *Semblance plot with t-v picks.*

In Figure 7.6, solid dots show where t-v picks were made. The vertical line labeled "Multiples" has a single velocity, 6800 f/s. Although events in the left side of Figure 7.7 are nicely flattened, this is misleading. We want multiples to stack out (attenuate), not reinforce. The NMO plot in the right side of Figure 7.7, the result of velan picks for primaries, shows poor event alignment below approximately 2.5 seconds. We see in Figure 7.7 that multiples cover up primaries. The best we can do is pick t-v pairs that will attenuate the multiples during stack. We do this so that after stack, the primaries will be easier to see (interpret).

Note: The mutes are different on the two data sets in Figure 7.7 because **sunmo** parameter *smute* affected the two moveout curves differently.

Table 7.3: *Time-Velocity picks on CMP oz14h.su*

Multipes	Primaries
tnmo=0.010,1.280,3.980	tnmo=0.020,1.590,2.520,3.97
vnmo=5000,6800,6800	vnmo=5000,7180,7980,10200

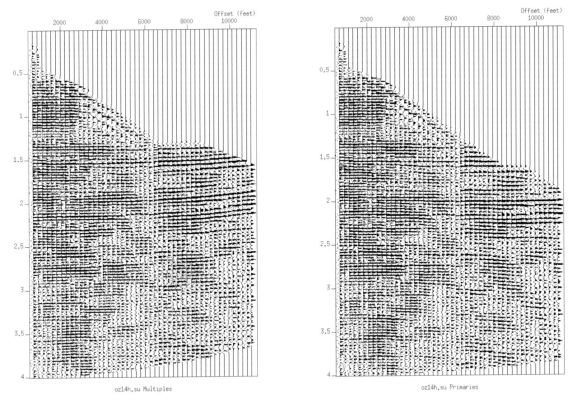

Figure 7.7: *Left: oz14h.su after NMO for multiples. Right: oz14h.su after NMO for primaries.*

7.6.3 Constant Velocity Stacks

Another velocity analysis plot is constant velocity stacks (CVS). Figure 7.8 is the result of making a CVS from CMP 60 of Model 4 (Figure 7.5). Figure 7.8 shows ten CVS panels. Each panel has 11 traces — always an odd number. (Individual traces within panels are difficult to see because we used **suximage** instead of **suxwigb**.)

The middle trace of each panel is CMP 60. The first CVS panel, the first 11 traces, is the result of applying NMO and stack to CMPs 55 to 65. This is also true for the next nine panels. (CVS plots cannot be made for a single CMP; only for a line of CMPs.) The NMO velocity for the first CVS panel is 1200 m/s. The NMO velocity for the second CVS panel is 1500 m/s. To make Figure 7.8, NMO and stack was done 110 times!

Each panel gives a simplified view of the final stacked section. The more continuity the events show and the higher their amplitudes, the better will be the final stacked result.

Because each panel uses only one velocity, the best picks over time usually do not fall within one single panel. (**Note:** This can be different for multi velocity stacks (MVS). Each panel of an MVS represents a time-varying velocity function. MVS are common in the industry today, but the idea behind CVS is not much different and it is easier to create in SU.) The first reflector has the highest amplitude (is best focused) near 1500 m/s, the second reflector has the highest amplitude near 2700 m/s, and the third reflector has the highest amplitude near 3000 m/s.

The arrows in Figure 7.8 emphasize that when we interpret the best velocity from each panel, we use the velocity above the center trace of each panel. The other traces in the panel provide visual emphasis to the best NMO velocity, but the center CMP of each panel is the one we study.

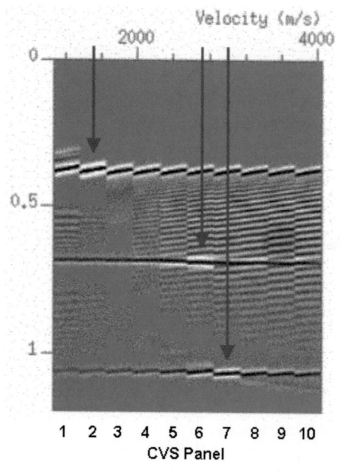

Figure 7.8: *A constant velocity stack (CVS) of 10 panels made from CMP 60, Model 4.*

Also, remember that the velocity spacing between panels (here) is 300 m/s. We cannot determine a precise stacking velocity from the CVS plot, but it is valuable when used with the velan. We will show this in a later section when we combine the original CMP display, the velan, and the CVS plot in an interactive velocity analysis script.

7.6.4 Quality Control (QC) after Picking

The semblance plot and the CVS are tools we use to decide which stacking velocities are reasonable – they provide a possibility for a first velocity pick. However, QC of the picked velocities afterwards is essential. The following interactive velocity analysis script provides three QC tools:

- a graph of the velocity profile,
- a plot of the NMO-corrected CMP, and
- a plot of the stack trace repeated eight times.

The velocity profile should be smooth and it should show generally increasing velocities with depth. While NMO-velocity inversions (decreases) are possible, those picks should be double-checked. The velocity profile should also show a pick near zero time and a pick near the last time. This last requirement ensures that when **sunmo** interpolates velocity functions between CMPs, the interpolation is guided throughout the time section.

The corrected CMP should exhibit flat primary reflections. After NMO is applied, an upward curved primary reflection means the picked velocity is too slow (overcorrected) and a downward curved primary reflection means the picked velocity is to fast (under-corrected).

A fourth QC tool is provided when you decide to re-pick a CMP: the previous picks are placed on the velan. This allows you to easily use or avoid previous picks.

Finally, how frequently along the seismic line must you do velocity analysis? It depends on the complexity of the geology.

7.6.5 Computer Note 11: Interactive Scripts

Throughout this Primer, you will see that we have two methods to run an interactive session. The two methods do not affect how data are processed. We prefer the second method because we find it convenient to open a separate window for interactive processing.

Method 1:   Run a *.sh* script. For example,

```
iva.sh
```

When we use this method, the terminal window in which we execute this command becomes the interactive dialog window.

Note: When using this method, never use the ampersand (&). That is, never use the following command:

```
iva.sh &
```

As explained in Section 1.9.1, "&" puts the process in the background, something you do not want to do when you are trying to be interactive!

Method 2:   Run a *.scr* script. For example,

```
idecon.scr &
```

Here, the ampersand is optional, as explained in Section 1.9.

When we use this method, the *.scr* script opens an interactive dialog window and starts a *.sh* script. Below is the *iva.scr* script:

```
1 #! /bin/sh
2 # File: iva.scr
3 #        Run this script to start script iva.sh
4
5 xterm -geom 8x20+10+545 -e iva.sh
6
```

This script has two active lines. Line 1 invokes the shell. Line 5 opens the interactive window and starts the *iva.sh* script. In line 5, "80x20" specifies the size of the dialog window: 80 characters wide by 20 characters high. Also, "10+545" specifies the placement of the upper left corner of the dialog window: 10 pixels from the left side of the viewing area and 545 pixels from the top of the viewing area.

7.6.6 Velocity Analysis: Scripts iva.scr and iva.sh

The following sections examine iva in detail. We would be pleased if you read every sentence; however, you can use iva without doing so. We suggest the following as minimum reading: Section 7.6.6.2 (iva Displays) and Section 7.6.6.4 (iva.sh – User-supplied Values).

7.6.6.1   iva.scr

You can do velocity analysis by creating a velan for each selected CMP along a line (script *oz14velan.sh*, Section 7.6.2), then picking t-v pairs from the velan. After you have t-v pairs for those selected CMPs, you can put them into a script that does NMO for the line by interpolating between analyzed CMPs. However, if you do not like the flattened CMPs or the later stack section, this style of velocity analysis is tedious to repeat. Script *iva.sh* allows you to make picks interactively, view the flattened CMPs, and if necessary, refine your picks.

As we explained in the previous section, we start interactive scripts by running a *.scr* script. Here is *iva.scr*:

```
1   #! /bin/sh
2   # File: iva.scr
3   #        Run this script to start script iva.sh
4
5   xterm -geom 80x20+10+545 -e iva.sh
6
```

Line 1 invokes the shell. Line 5 opens a dialog window of specified size and position. Line 5 also starts *iva.sh*, the processing script, within that window.

7.6.6.2   iva Displays

Before looking at *iva.sh*, let's look at the displays it creates. For the discussion in this section, we limit processing to two CMPs: CMP 60 and CMP 80.

We start the interactive velocity analysis script by entering:

iva.scr &

The script prints the following information in the dialog window (as well as other system messages that we do not show) and creates the plots shown in Figure 7.9.

```
*** INTERACTIVE VELOCITY ANALYSIS ***


Preparing CMP 1 of 2 for Picking
Location is CMP 60
  Start CVS CMP = 55     End CVS CMP = 65

  Use the semblance plot to pick (t,v) pairs.
  Type "s" when the mouse pointer is where you want a pick.
  Be sure your picks increase in time.
  To control velocity interpolation, pick a first value
    near zero time and a last value near the last time.
  Type "q" in the semblance plot when you finish picking.
```

Following the directions above, we click on the border of the velan box to activate it, then we make a series of picks in the velan image:

- Put the mouse pointer over a time near zero and a velocity near that of water (because we know the top layer in the model is water (velocity = 1500 m/s), then press letter "S."

- Put the mouse pointer over the first major event, at approximately 0.375 seconds and velocity approximately 1500 m/s, then press letter "S."

- Put the mouse pointer over the second major event, at approximately 0.7 seconds and velocity approximately 2700 m/s, then press letter "S."

- Put the mouse pointer over the third major event, at approximately 1.09 seconds and velocity approximately 2900 m/s, then press letter "S."

- You picked all the major events. Now make a final pick near the end time: Following the t-v trend of the last two picks, put the mouse pointer near the last time, 2 seconds, then press letter "S."

- We are finished picking t-v pairs, so we press letter "Q."

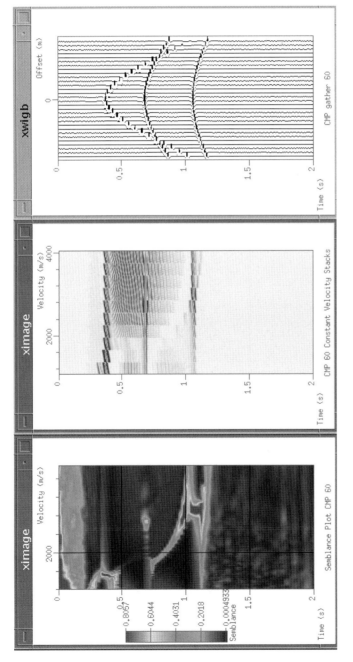

Figure 7.9: *Left: The velan. Middle: The CVS plot. Right: A plot of the input CMP.*

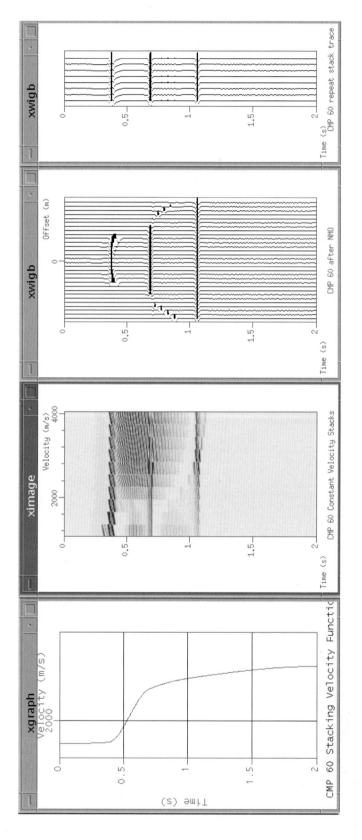

Figure 7.10: *After picking. Left: The t-v graph. Middle-left: The CVS plot. Middle-right: The CMP after NMO. Right: The stack trace repeated eight times.*

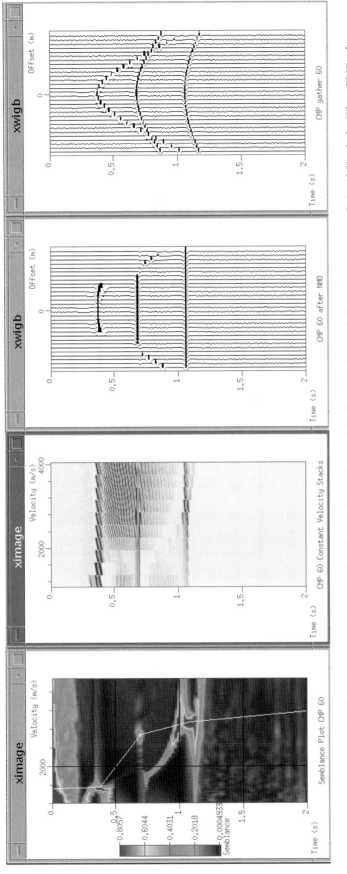

Figure 7.11: *The display for re-picking. Left: The color velan with the previous picks superimposed. Middle-left: The CVS plot. Middle-right: The CMP after NMO. Right: A plot of the input CMP.*

The other plots are not just for show. At any time during picking, you can click the middle mouse button in any window to get information. For example, by clicking near the apex of the third reflector in the CMP window (Figure 7.12 below), the time, 1.06596 seconds tells you where to expect the third reflector to show maximum semblance in the velan. You can then use the middle mouse button to click in the velan window to find the semblance maximum that corresponds to 1.06596 seconds. (Are the numbers really accurate to five decimal places? No.)

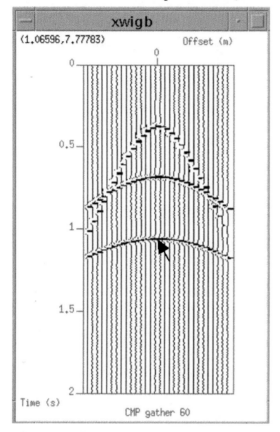 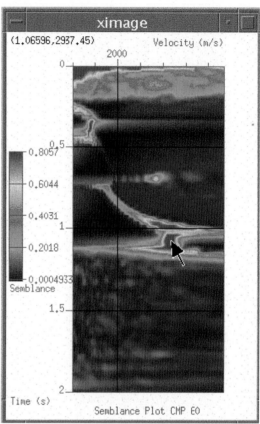

Figure 7.12: *The upper left corner of a plot shows y,x information when you click the middle mouse button in a window. Left: In the CMP window, y,x corresponds to time,offset. Right: In the velan window, y,x corresponds to time,velocity.*

Note: Whenever an **suximage** window is active, you can scroll through color palettes by repeatedly pressing "r," "R," "h," or "H.". The following are lines from the **ximage** selfdoc. (RGB stands for Red, Green, Blue; HSV stands for Hue, Saturation, Value.)

```
X Functionality:
 Button 1        Zoom with rubberband box
 Button 2        Show mouse (x1,x2) coordinates while pressed
 q or Q key      Quit
 s key           Save current mouse (x1,x2) location to file
 p or P key      Plot current window with pswigb (only from disk files)

... change colormap interactively
 r              install next RGB - colormap
 R              install previous RGB - colormap
 h              install next HSV - colormap
 H              install previous HSV - colormap
 H              install previous HSV - colormap
```

```
(Move mouse cursor out and back into window for r,R,h,H to take effect)
```

Once "Q" is pressed when the velan window is active, the plots of Figure 7.10 appear. The t-v graph looks good because (1) it starts at the top of the graph and ends near the bottom, and (2) it is smooth, and (3) it generally increases with depth.

The middle-right plot, the CMP after NMO, is also pleasing because it shows flat reflectors.

The right plot, the repeat stack, shows strong amplitude peaks for the reflectors.

The dialog window shows the following:

```
t-v PICKS CMP 60
--------------------
0.0211082  1526.7
0.37467    1526.7
0.69657    2685
1.08707    2952.3
1.98945    3115.65

  Use the velocity profile (left),
    the NMO-corrected gather (middle-right),
    and the repeated stack trace (right)
    to decide whether to re-pick the CMP.

Picks OK? (y/n)
```

The dialog window list of t-v picks confirms that we have a pick near zero time (0.0211082 seconds) and a pick near the end time (1.98945 seconds).

The picks are good, but we want to see the re-picking plots. So, we press "N" then the Enter or Return key. The dialog window shows the following:

```
Picks OK? (y/n)
n

Repick CMP 60. Overlay previous picks.

Preparing CMP 1 of 2 for Picking
Location is CMP 60
  Start CVS CMP = 55    End CVS CMP = 65

  Use the semblance plot to pick (t,v) pairs.
  Type "s" when the mouse pointer is where you want a pick.
  Be sure your picks increase in time.
  To control velocity interpolation, pick a first value
    near zero time and a last value near the last time.
  Type "q" in the semblance plot when you finish picking.
```

Figure 7.11 shows that the velan re-appears with an overlay of the picks. Also, the original CMP appears on the right, covering the repeat stack plot.

We repeat our picks, guided by the white line. When we are finished, we press "Q." Plots similar to Figure 10 appear and the dialog window shows:

```
t-v PICKS CMP 60
--------------------
0.0158311  1497
0.369393   1526.7
0.69657    2670.15
1.08707    2952.3
1.98945    3130.5

  Use the velocity profile (left),
    the NMO-corrected gather (middle-right),
```

```
and the repeated stack trace (right)
to decide whether to re-pick the CMP.
```

```
Picks OK? (y/n)
```

We are finished picking CMP 60, so we can enter "Y" or merely press Enter or Return. (You will see this below in script *iva.sh*.) We press Return. The dialog window shows:

```
-- CLOSING CMP 60 WINDOWS --

Preparing CMP 2 of 2 for Picking
Location is CMP 80
  Start CVS CMP = 75    End CVS CMP = 85

  Use the semblance plot to pick (t,v) pairs.
  Type "s" when the mouse pointer is where you want a pick.
  Be sure your picks increase in time.
  To control velocity interpolation, pick a first value
    near zero time and a last value near the last time.
  Type "q" in the semblance plot when you finish picking.
```

After a CMP is picked and before the plots for the next CMP are created, all plot windows are closed.

Notice that a new series of CMPs is used to create the CVS plot for CMP 80.

After we "pick" CMP 80, and approve the picks (enter "Y" or merely press Enter or Return when asked "Picks OK? (y/n)"), the dialog window shows:

```
-- CLOSING CMP 80 WINDOWS --

The output file of t-v pairs is vpick.txt
```

```
press return key to continue
```

After pressing Return or Enter, the dialog window closes.

Note: If, after a second try, we did not like our picks, we would repeat our "N" answer to "Picks OK? (y/n)" and we would see plots as shown in Figure 11. This time the velan would have an overlay of our second set of picks.

Note: The output file of t-v picks is called "vpick.txt." We recommend that you immediately change the name of this file so it will not be overwritten if you re-run *iva.scr*. Alternatively, you can change the name of the "outpicks" file in script *iva.sh*.

Here is file *vpicks.txt*:

```
cdp=60,80
#=1,2
tnmo=0.0158311,0.369393,0.69657,1.08707,1.98945
vnmo=1497,1526.7,2670.15,2952.3,3130.5
tnmo=0.0158311,0.353562,0.69657,1.05541,1.98945
vnmo=1497,1541.55,2655.3,2774.1,2878.05
```

After minor editing, this file can be used by **sunmo**.

7.6.6.3   iva.sh — Introduction

Script *iva.sh* has the following major sections:

- Lines 8-47 are user-supplied values.

- Lines 48-83 document semblance (velan) and CVS computations. This section includes computing the last semblance velocity (parameter *lvs*, lines 57-59).

- Lines 84-118 document files created by *iva.sh*.

- Lines 119-159 do some one-time processing. Old temporary files are removed in line 127. Old files can be left on the system when we crash the script. Remember, you can terminate the script at any time by pressing "Control-C." If you do not plan to re-run the script right away, you can remove the temporary files by entering a command similar to line 127:

      rm panel.* picks.* par.* tmp*

  Line 158 sets variable *repick* to "false" since, at this point, the user has not yet made a first pick.

- Lines 160-440 are the core processing and plotting.

- Lines 441-453 create the velocity picks output file (*outpicks*).

- Lines 454-461 write a message to the screen telling us the name of the output file of picks, **pause** the script, remove all temporary files, and close the script. The **pause** of line 459 also gives us the chance to examine the temporary files before they are removed.

```
1   #! /bin/sh
2   # File: iva.sh
3   #        Run script iva.scr to start this script
4
5   # Set messages on
6   ##set -x
7
8   #================================================
9   # USER AREA -- SUPPLY VALUES
10  #------------------------------------------------
11  # CMPs for analysis
12
13  cmp1=60   cmp2=80   cmp3=100
14  cmp4=120  cmp5=140  cmp6=160
15  cmp7=180  cmp8=200  cmp9=220
16  cmp10=240 cmp11=260 cmp12=280
17  cmp13=300 cmp14=320 cmp15=340
18  cmp16=360 cmp17=380 cmp18=400
19
20  numCMPs=18
21
22  #------------------------------------------------
23  # File names
24
25  indata=cmp4.su       # SU format
26  outpicks=vpick.txt   # ASCII file
27
28  #------------------------------------------------
29  # display choices
30
31  myperc=98        # perc value for plot
32  plottype=0       # 0 = wiggle plot,  1 = image plot
33
34  #------------------------------------------------
35  # Processing variables
36
37  # Semblance variables
38  nvs=100  # number of velocities
39  dvs=27   # velocity intervals
40  fvs=1200 # first velocity
41
```

```
42   # CVS variables
43   fc=1200 # first CVS velocity
44   lc=3900 # last CVS velocity
45   nc=10   # number of CVS velocities (panels)
46   XX=11   # ODD number of CMPs to stack into central CVS
47
48   #====================================================
49
50   # HOW SEMBLANCE (VELAN) VELOCITIES ARE COMPUTED
51
52   # Last Vel =  fvs + (( nvs-1 ) * dvs ) = lvs
53   #     5000 =  500 + ((  99-1 ) * 45  )
54   #     3900 = 1200 + (( 100-1 ) * 27  )
55
56   # Compute last semblance (velan) velocity
57   lvs=`bc -l << -END
58   $fvs + (( $nvs - 1 ) * $dvs )
59   END`
60
61   #-------------------------------------------------
62
63   # HOW CVS VELOCITIES ARE COMPUTED
64
65   # dc = CVS velocity increment
66   # dc = ( last CVS vel - first CVS vel ) / ( # CVS - 1 )
67   # m = CVS plot trace spacing (m = d2, vel units)
68   # m = ( last CVS vel - first CVS vel ) / ( ( # CVS - 1 ) * XX )
69
70   # j=1
71   # while [ j le nc ]
72   # do
73   #   vel = fc  + { [(   lc - fc   ) / ( nc-1 )] * ( j-1) }
74   #   j = j + 1
75   # done
76   # EXAMPLE:
77   #   vel = 1200 + ( (( 3900 - 1200 ) / ( 10-1 )) * ( 1-1) )
78   #   vel = 1200 + ( (( 3900 - 1200 ) / ( 10-1 )) * ( 2-1) )
79   #                                                .
80   #                                                .
81   #                                                .
82   #   vel = 1200 + ( (( 3900 - 1200 ) / ( 10-1 )) * (11-1) )
83
84   #====================================================
85
86   # FILE DESCRIPTIONS
87
88   # tmp0 = binary temp file for input CVS gathers
89   # tmp1 = binary temp file for output CVS traces
90   # tmp2 = ASCII temp file for managing picks
91   # tmp3 = binary temp file for stacked traces
92   # tmp4 = ASCII temp file for "wc" result (velan)
93   # tmp5 = ASCII temp file for stripping file name from tmp4 (velan)
94   # tmp6 = ASCII temp file to avoid screen display of "zap"
95   # tmp7 = ASCII temp file for picks
96   # tmp8 = binary temp file for NMO (flattened) section
97   # panel.$picknow = current CMP windowed from line of CMPs
98   # picks.$picknow = current CMP picks arranged as "t1 v1"
99   #                                                "t2 v2"
100  #                                                etc.
101  # par.# (# is a sequential index number; 1, 2, etc.)
102  #        = current CMP picks arranged as
103  #          "tnmo=t1,t2,t3,...
104  #          "vnmo=v1,v2,v3,...
```

```
105  # par.uni.#  (# is a sequential index number; 1, 2, etc.)
106  #       = current CMP picks arranged as
107  #           "xin=t1,t2,t3,..."
108  #           "yin=v1,v2,v3,..."
109  #           for input to xgraph to display velocity profile
110  # par.cmp = file of CMP number and sequential index number;
111  #           for example:  "40 1"
112  #                         "60 2"
113  #                          etc.
114  # par.0 = file "par.cmp" re-arranged as
115  #           "cdp=#,#,#,etc."  NOTE: # in this line is picked CMP
116  #           "#=1,2,3,etc."    NOTE: # in this line is "#"
117  # outpicks = concatenation of par.0 and all par.# files.
118
119  #================================================
120
121  echo " "
122  echo "  *** INTERACTIVE VELOCITY ANALYSIS ***"
123  echo " "
124
125  #-------------------------------------------------
126  # Remove old files.  Open new files
127  rm -f panel.* picks.* par.* tmp*
128
129  > $outpicks  # Write empty file for final picks
130  > par.cmp    # Write empty file for recording CMP values
131
132  #-------------------------------------------------
133  # Get ns, dt, first time from seismic file
134  nt=`sugethw ns < $indata | sed 1q | sed 's/.*ns=//'`
135  dt=`sugethw dt < $indata | sed 1q | sed 's/.*dt=//'`
136  ft=`sugethw delrt < $indata | sed 1q | sed 's/.*delrt=//'`
137
138  # Convert dt from header value in microseconds
139  # to seconds for velocity profile plot
140  dt=`bc -l << -END
141    scale=6
142    $dt / 1000000
143  END`
144
145  # XGRAPH: If "delrt", use it; else use zero
146  if [ $ft -ne 0 ] ; then
147    tstart=`bc -l << -END
148      scale=6
149      $ft / 1000
150    END`
151  else
152    tstart=0.0
153  fi
154
155  #-------------------------------------------------
156
157  # Initialize "repick" -- for plotting previous picks on velan
158  repick=1  # 1=false, 0=true
159
160  #-------------------------------------------------
161  # BEGIN IVA LOOP
162  #-------------------------------------------------
163
164  i=1
165  while [ $i -le $numCMPs ]
166  do
167
```

```
168   # set variable $picknow to current CMP
169     eval picknow=\$cmp$i
170
171     if [ $repick -eq 1 ] ; then
172       echo " "
173       echo "Preparing CMP $i of $numCMPs for Picking "
174       echo "Location is CMP $picknow "
175     fi
176
177   #--------------------------------------------------
178   # Plot CMP (right)
179   #--------------------------------------------------
180
181     suwind < $indata \
182             key=cdp min=$picknow max=$picknow > panel.$picknow
183     if [ $repick -eq 1 ] ; then
184       if [ $plottype -eq 0 ] ; then
185         suxwigb < panel.$picknow xbox=634 ybox=10 wbox=300 hbox=450 \
186                   title="CMP gather $picknow" \
187                   label1=" Time (s)" label2="Offset (m)" key=offset \
188                   perc=$myperc verbose=0 &
189       else
190         suximage < panel.$picknow xbox=634 ybox=10 wbox=300 hbox=450 \
191                   title="CMP gather $picknow" \
192                   label1=" Time (s)" \
193                   perc=$myperc verbose=0 &
194       fi
195     else
196       if [ $plottype -eq 0 ] ; then
197         suxwigb < panel.$picknow xbox=946 ybox=10 wbox=300 hbox=450 \
198                   title="CMP gather $picknow" \
199                   label1=" Time (s)" label2="Offset (m)" key=offset \
200                   perc=$myperc verbose=0 &
201       else
202         suximage < panel.$picknow xbox=946 ybox=10 wbox=300 hbox=450 \
203                   title="CMP gather $picknow" \
204                   label1=" Time (s)" \
205                   perc=$myperc verbose=0 &
206       fi
207     fi
208
209   #--------------------------------------------------
210   # Constant Velocity Stacks (CVS) (middle-left)
211   # Make CVS plot for first pick effort.
212   # If re-picking t-v values, do not make this plot.
213   #--------------------------------------------------
214
215   # repick: 1=false, 0=true
216     if [ $repick -eq 1 ] ; then
217
218   # number of CMPs - 1; for windowing
219       X=`expr $XX - 1`
220
221   # Window CMPs around central CMP (+/- X/2). Write to tmp0
222       k1=`expr $picknow - $X / 2`  # Window from CMP to CMP - X/2
223       k2=`expr $picknow + $X / 2`  # Window from CMP to CMP + X/2
224       suwind < $indata key=cdp min=$k1 max=$k2 > tmp0
225
226   # Calculate CVS velocity increment
227   # dc = ( last CVS vel - first CVS vel ) / ( # CVS - 1 )
228       dc=`bc -l << -END
229       ( $lc - $fc ) / ( $nc - 1 )
230       END`
```

```
231
232   # Calculate trace spacing for CVS plot (m = d2, vel units)
233   # m = ( last CVS vel - first CVS vel ) / ( ( # CVS - 1 ) * XX )
234       m=`bc -l << -END
235       ( $lc - $fc ) / ( ( $nc - 1 ) * $XX )
236       END`
237
238   # CVS velocity loop
239       j=1
240       while [ $j -le $nc ]
241       do
242
243         vel=`bc -l << -END
244         $fc + $dc * ( $j - 1 )
245         END`
246
247   # uncomment to print CVS velocities to screen
248   ##      echo " vel = $vel"
249
250         sunmo < tmp0 vnmo=$vel verbose=0 |
251         sustack >> tmp1
252
253         j=`expr $j + 1`
254       done
255
256   # Compute lowest velocity for annotating CVS plot
257   # loV = first CVS velocity - ( ( CMP range - 1 ) / 2 ) * vel inc
258       loV=`bc -l << -END
259       $fc - ( $X / 2) * $m
260       END`
261
262       suximage < tmp1 xbox=322 ybox=10 wbox=300 hbox=450 \
263                 title="CMP $picknow Constant Velocity Stacks" \
264                 label1=" Time (s)" label2="Velocity (m/s)" \
265                 f2=$loV d2=$m verbose=0 \
266                 perc=$myperc n2tic=5 cmap=rgb0 &
267
268     fi
269
270   #-------------------------------------------------
271   # Picking instructions
272   #-------------------------------------------------
273
274     echo " "
275     echo "Preparing CMP $i of $numCMPs for Picking "
276     echo "Location is CMP $picknow "
277     echo "  Start CVS CMP = $k1    End CVS CMP = $k2"
278     echo " "
279     echo "  Use the semblance plot to pick (t,v) pairs."
280     echo "  Type \"s\" when the mouse pointer is where you want a pick."
281     echo "  Be sure your picks increase in time."
282     echo "  To control velocity interpolation, pick a first value"
283     echo "    near zero time and a last value near the last time."
284     echo "  Type \"q\" in the semblance plot when you finish picking."
285
286   #-------------------------------------------------
287   # Plot semblance (velan) (left)
288   #-------------------------------------------------
289
290   # repick: 1=false, 0=true
291     if [ $repick -eq 0 ] ; then
292
293   #--- --- --- --- --- --- --- --- --- ---
```

```
294  # Get the number of picks (number of lines) in tmp7 |
295  #    Remove blank spaces preceding the line count.
296  # Remove file name that was returned from "wc".
297  # Store line count in "npair" to guide line on velan.
298
299      wc -l tmp7 | sed 's/^  *\(.*\)/\1/' > tmp4
300      sed 's/tmp7//' < tmp4 > tmp5
301      npair=`sort < tmp5`
302  #---  ---  ---  ---  ---  ---  ---  ---  ---  ---
303
304      suvelan < panel.$picknow nv=$nvs dv=$dvs fv=$fvs |
305      suximage xbox=10 ybox=10 wbox=300 hbox=450 perc=99 \
306              units="semblance" f2=$fvs d2=$dvs n2tic=5 \
307              title="Semblance Plot CMP $picknow" cmap=hsv2 \
308              label1=" Time (s)" label2="Velocity (m/s)" \
309              legend=1 units=Semblance verbose=0 gridcolor=black \
310              grid1=solid grid2=solid mpicks=picks.$picknow \
311              curve=tmp7 npair=$npair curvecolor=white
312
313    else
314
315      suvelan < panel.$picknow nv=$nvs dv=$dvs fv=$fvs |
316      suximage xbox=10 ybox=10 wbox=300 hbox=450 perc=99 \
317              units="semblance" f2=$fvs d2=$dvs n2tic=5 \
318              title="Semblance Plot CMP $picknow" cmap=hsv2 \
319              label1=" Time (s)" label2="Velocity (m/s)" \
320              legend=1 units=Semblance verbose=0 gridcolor=black \
321              grid1=solid grid2=solid mpicks=picks.$picknow
322
323    fi
324
325  #------------------------------------------------
326  # End first set of plots
327  #================================================
328
329  #------------------------------------------------
330  # Manage picks (1): Prepare picks for sunmo
331  #------------------------------------------------
332
333    sort < picks.$picknow -n |
334      mkparfile string1=tnmo string2=vnmo > par.$i
335    echo "cdp=$picknow" >> tmp2
336    cat par.$i >> tmp2
337
338  #================================================
339  # Begin second set of plots
340  #------------------------------------------------
341
342  #------------------------------------------------
343  # Flattened seismic data (NMO) plot (middle-right)
344  #------------------------------------------------
345
346    sunmo < panel.$picknow par=tmp2 verbose=0 > tmp8
347    if [ $plottype -eq 0 ] ; then
348      suxwigb < tmp8 xbox=634 ybox=10 wbox=300 hbox=450 \
349              title="CMP $picknow after NMO" \
350              label1=" Time (s)" label2="Offset (m)" \
351              verbose=0 perc=$myperc key=offset &
352    else
353      suximage < tmp8 xbox=634 ybox=10 wbox=300 hbox=450 \
354              title="CMP $picknow after NMO" \
355              label1=" Time (s)" \
356              verbose=0 perc=$myperc &
```

```
357    fi
358
359  #---------------------------------------------------
360  # Stack window (right)
361  #---------------------------------------------------
362
363    j=1
364    while [ $j -le 8 ]
365    do
366
367  # Append stack trace into tmp3 multiple times
368      sustack < tmp8 >> tmp3
369
370      j=`expr $j + 1`
371    done
372
373    suxwigb < tmp3 xbox=946 ybox=10 wbox=200 hbox=450 \
374            title="CMP $picknow repeat stack trace" \
375            label1=" Time (s)" d2num=50 key=cdp \
376            verbose=0 perc=$myperc &
377
378  #----------------------------------------------------
379  # Manage picks (2): Prepare picks for vel profile
380  #----------------------------------------------------
381
382    sed < par.$i '
383    s/tnmo/xin/
384    s/vnmo/yin/
385                ' > par.uni.$i
386
387  #----------------------------------------------------
388  # Velocity profile (left)
389  #----------------------------------------------------
390
391    unisam nout=$nt fxout=$tstart dxout=$dt \
392            par=par.uni.$i method=mono |
393    xgraph n=$nt nplot=1 d1=$dt f1=$tstart x2beg=$fvs x2end=$lvs \
394            label1=" Time (s)" label2="Velocity (m/s)" \
395            title="CMP $picknow Stacking Velocity Function" \
396            -geometry 300x450+10+10 -bg white style=seismic \
397            grid1=solid grid2=solid linecolor=2 marksize=1 mark=0 \
398            titleColor=black axesColor=blue &
399
400  #----------------------------------------------------
401  # Dialogue with user: re-pick ?
402  #----------------------------------------------------
403
404    echo " "
405    echo " t-v PICKS CMP $picknow"
406    echo "--------------------"
407    cat picks.$picknow
408    echo " "
409    echo "  Use the velocity profile (left),"
410    echo "    the NMO-corrected gather (middle-right),"
411    echo "    and the repeated stack trace (right)"
412    echo "    to decide whether to re-pick the CMP."
413    echo " "
414    echo "Picks OK? (y/n) " > /dev/tty
415    read response
416
417    rm tmp*
418
419  # "n" means re-loop. Otherwise, continue to next CMP.
```

```
420    case $response in
421      n*)
422          i=$i
423          echo " "
424          echo "Repick CMP $picknow. Overlay previous picks."
425          repick=0
426          cp picks.$picknow tmp7
427          ;;
428      *)
429          echo "$picknow  $i" >> par.cmp
430          i=`expr $i + 1`
431          repick=1
432          echo "-- CLOSING CMP $picknow WINDOWS --"
433          zap xwigb > tmp6
434          zap ximage > tmp6
435          zap xgraph > tmp6
436          ;;
437    esac
438
439  done
440
441  #-------------------------------------------------
442  # Create velocity output file
443  #-------------------------------------------------
444
445  mkparfile < par.cmp string1=cdp string2=# > par.0
446
447  i=0
448  while [ $i -le $numCMPs ]
449  do
450    sed < par.$i 's/$/ \\/g' >> $outpicks
451    i=`expr $i + 1`
452  done
453
454  #-------------------------------------------------
455  # Remove files and exit
456  #-------------------------------------------------
457  echo " "
458  echo " The output file of t-v pairs is "$outpicks
459  pause
460  rm -f panel.* picks.* par.* tmp*
461  exit
462
```

7.6.6.4   iva.sh — User-supplied Values

Line 20 has the number of CMPs to process, the number of times to loop through velocity analysis. Line 20 is tied to the velocity analysis loop with the following lines:

```
 20   numCMPs=18
164   i=1
165   while [ $i -le $numCMPs ]
166   do
430          i=`expr $i + 1`
439   done
```

Line 164 initializes counter variable "*i*"; line 430 increments counter "*i*" when we finish picking a CMP and want to move to the next CMP or quit the script. Line 165 ensures that only the number of CMPs in line 20 are processed.

Note: In line 20, you can use any value up to the total number of CMPs you list. We sometimes use a value of "2" or "3" when we are testing the script or previewing the data.

Lines 13-18 list the CMPs. Here, the CMPs evenly increment by 20, but any values are acceptable. The CMPs are listed on six lines, but they can be listed on three lines, two lines, or even one per line to use 18 lines. The number per line is not important. However, the style of these lines is very important. When the **eval** function operates, line 169, it sets the value of *picknow*, the current CMP.

```
169     eval picknow=\$cmp$i
```

The first time line 169 is used, the value of *picknow* becomes the value of parameter *cmp1*. The script searches its memory and finds that `cmp1=60`, so another part of the script (lines 181-182) windows CMP 60 from the input file. The second time line 156 is used, the value of *picknow* becomes the value of parameter *cmp2*. The script searches its memory and finds that `cmp2=80`, so lines 181-182 extract CMP 80 from the input file. Etc. Because line 169 uses "cmp," lines 13-18 must also use "cmp."

Note: The Bourne shell does not have "tables." To get around this restriction, lines 13-18 list the table elements and line 20 has the number of table elements. Line 169 concatenates the *$cmp* variable with the value of the *$i* variable. Line 169 effectively selects the appropriate array element. This pseudo-table structure allows us to specify unequally spaced CMPs. We thank Brian Weeden for helping us understand **eval**.

Line 26 is the place for the name of the output t-v pick file. We suggest that if you do not rename this file here with every run, that you rename the output file after each run.

Before you set the *perc* value in line 31, we suggest you determine a *perc* value by using **suxwigb** or **suximage** externally.

Note: You might want to have a separate *perc* value for the occurrences of **suximage** that follow **suvelan** (lines 304-321). For example, after line 31, add the following line:

```
percvelan=98    # perc value for velan ximage
```

Then, on lines 305 and 316, change `perc=99` to `perc=$percvelan`. Using *percvelan*, you can change the "amplitude" of the velan plots separately from the other plots. This is a powerful tool for revealing details in real data.

Line 32 lets you set the CMP plot and the flattened CMP plot as wiggle trace plots or image plots. This option is here because we find that "too many" traces do not display well in **suxwigb**. ("Too many" is qualitative!)

Lines 37-46 let you set the semblance and CVS variables. To understand the semblance variables, use the documentation on lines 52-54, study lines 304 and 315, and read the **suvelan** selfdoc. To understand the CVS variables, use the documentation on lines 63-82 and study lines 218-268.

Note: You can uncomment line 248 to print the CVS velocities to the screen (see Section 7.6.3).

7.6.6.5   iva.sh — File Documentation

Lines 84-118 describe the files created by *iva.sh*. The small files are ASCII and the large ones contain seismic data. During processing, you can investigate the contents of the ASCII files by using **cat** and you can investigate the contents of the seismic files by using **suxwigb** or **suximage**.

7.6.6.6 iva.sh — One-time Processing

Line 127 removes old temporary files if a previous run that crashed did not permit execution of line 460.

```
127   rm -f panel.* picks.* par.* tmp*
460   rm -f panel.* picks.* par.* tmp*
```

Line 129 opens the final output file. We open the file here to make sure the command works instead of waiting until all processing is finished.

```
129   > $outpicks  # Write empty file for final picks
```

Line 134 extracts the value of key *ns* from the first trace of the input file; line 135 extracts the value of key *dt* from the first trace of the input file. We use these values in lines 391 and 393 to create the velocity profile plot.

```
134   nt=`sugethw ns < $indata | sed 1q | sed 's/.*ns=//'`
135   dt=`sugethw dt < $indata | sed 1q | sed 's/.*dt=//'`
```

Every trace has the same value for keys *ns* and *dt* (see the **surange** output of *cmp4.su*, Section 7.2). The syntax of lines 134 and 135 can be used to extract the value of any key from the first trace of the input seismic data.

Line 141, `scale=6`, sets the number of significant figures (6) of the computed *dt* value, lines 140-143.

```
140   dt=`bc -l << -END
141      scale=6
142      $dt / 1000000
143   END`
```

Line 136 extracts the value of key *delrt*: delay recording time (milliseconds). As with *ns* and *dt*, this is used for the velocity profile. If the input file was windowed in time, the non-zero value of *delrt* is converted to seconds (lines 147-150) and used as parameter *tstart* to label the beginning time of the velocity profile (lines 391-398). If *delrt* is not used or contains zero, *tstart* is directly set to zero.

Because we check the value of *delrt*, and, if it is non-zero, we use it to set the start time of the velocity profile, we can do velocity analysis on a time-windowed data set. However, as was pointed out in Section 7.6.4, "The velocity profile should also show a pick near zero time and a pick near the last time." If a time-windowed file is used, after picking, the pick file must be edited to include picks near zero time and near the last times.

7.6.6.7 iva.sh — IVA Loop

This is a big section, so we examine it in blocks.

Lines 181-182 window a CMP from the input file.

If this is the first time we see the CMP; that is, this is not a re-pick (*repick*=1), the upper left corner of the CMP plot is placed at x=634, y=10 (line 185 or 190). If we are viewing the CMP after we picked t-v pairs, the CMP plot is placed farther to the right so it does not cover the flattened (NMO) plot. (The flattened CMP plot is created and plotted in lines 346-357.)

To make a CVS panel, a single velocity flattens an odd number of CMPs, then each CMP is stacked, then all the stack traces are displayed side-by-side. Line 46 (variable *XX*) specifies that eleven CMPs are to be used for each panel. Line 45 (variable *nc*) specifies the number of single velocities to apply to those eleven CMPs; in other words, the number of panels.

```
45   nc=10    # number of CVS velocities (panels)
46   XX=11    # ODD number of CMPs to stack into central CVS
```

If the first CMP is 60, CMP 60 is the panel's center CMP. Lines 218-223 compute the minimum and maximum CMP values of this series of eleven CMPs. Line 224 windows the eleven CMPs from the input file and puts them in file *tmp0*. In lines 239-254, as each of the ten velocities is computed, NMO is applied to *tmp0* and the CMPs are stacked. Line 251 uses the append command to concatenate successive panels into file *tmp1*.

```
251         sustack >> tmp1
```

File, *tmp1* is plotted by **suximage**, lines 262-266. The CVS horizontal axis is velocity. Lines 234-236 compute velocity trace spacing (*m*) and lines 258-260 compute the first (lowest) annotation velocity (*loV*)for the CVS **suximage** plot (line 265).

Lines 274-284 print the picking instructions to the screen. They also print information about the CMP that is about to be displayed (variables *numCMPs*, *picknow*, *k1*, and *k2* in lines 275-277). Notice that lines 173-174 previously printed this information. We repeat the information because system messages printed when **sunmo** runs scrolls earlier text out of the viewing area.

Lines 315-321 create and plot the velan if this is the first time the velan is created for the current CMP. Lines 299-311 are used if the velan is being displayed for re-picking. In both cases, user-supplied values for *nvs*, *dvs*, and *fvs* are used (lines 304, 306, 315, 317). For re-picking, we have to tell **suximage**, in line 311, the name of the file of the previous picks (parameter *curve*) and how many t-v picks are in the file (parameter *npair*).

```
311         curve=tmp7 npair=$npair curvecolor=white
```

Why does file *tmp7* hold the picks? Here we explain how picks are made and how they get into *tmp7*. Notice that **suximage**, **suxwigb**, and **xgraph** always end with an ampersand (&) except the two times **suximage** is used for picking: lines 305-311 and lines 316-321. In these two uses of **suximage**, parameter *mpicks* is used and has the value *picks.$picknow*:

```
310         grid1=solid grid2=solid mpicks=picks.$picknow \
321         grid1=solid grid2=solid mpicks=picks.$picknow
```

Because these uses of **suximage** do not use the ampersand (&), script control goes to **ximage**. When the velan plot is active and you press "Q," control leaves **ximage** and returns to the script (see the **ximage** selfdoc).

Whether we are picking a CMP the first time, the second time, or … the picks go in the same file. However, every time we choose to re-pick (the re-pick yes/no **case** is lines 420-437), the old picks are copied to *tmp7* by line 426:

```
426        cp picks.$picknow tmp7
```

So, *tmp7* has the old picks to overlay on the velan while new picking is going on.

Now we return to our travel down the script.

Lines 299-301 put the number of t-v picks (the number of lines) of *tmp7* into variable *npair*. Unfortunately, command **wc –l** also outputs the file name. For example, **wc –l** with the *iva.sh* script returns:

```
·····449 iva.sh
```

Those dots show that the output line count (449) is preceded by five blank spaces. The second part of line 299, the sed command, removes the preceding blank spaces (Lamb, 1990, page 91, number 11). We know that the name of the t-v pick file is *tmp7*; we use that information on line 300 to remove the file name from *tmp4*. The **sort** on line 301 is merely a device to copy the contents of *tmp5* (the line count) to variable *npair*.

After picking (after "Q" is pressed), lines 333-336 create a parameter (par) file formatted for input to **sunmo**. Command **mkparfile** is an SU function that has a selfdoc. The following are the contents of *tmp2* after picking CMP 60:

```
cdp=60
tnmo=0.0158311,0.369393,0.69657,1.08707,1.98945
vnmo=1497,1526.7,2670.15,2952.3,3130.5
```

File *tmp2* is used by **sunmo** in line 346 to make the flattened CMP for a QC display. The flattened CMP, file *tmp8*, is used by the next QC display, the repeat stack trace (lines 363-376). You might want fewer or more traces than the eight specified in line 364.

Lines 382-392 create a par file for the QC t-v graph. While **sunmo** requires *tnmo-vnmo* pairs, **xgraph** requires *xin-yin* pairs (lines 383-384). SU program **unisam** creates a smooth line from a series of point pairs (lines 391-392).

By the time we arrive at line 415, we are viewing the t-v graph, the CVS plot, the flattened CMP, and the repeat stacks. Also, the dialog window shows the current t-v pairs (line 407). Here, we have to choose whether to re-pick the CMP or accept our picks. No matter which we choose, all files with the *tmp* prefix are removed at line 417 to clear the files for the next round of displays.

If we choose to re-pick, the old picks are copied to *tmp7* for overlay on the next velan (line 426). If we choose to accept the current picks, the CMP counter increments (line 430), and all the plot windows are closed by lines 433-435. These lines direct (>) the command **zap** to file *tmp6* in order to minimize system messages to the screen.

7.6.6.8   iva.sh — Output the Picks

At the finish of processing, file *par.0* is created (line 445). Then (lines 447-452), file *par.0* is concatenated with files par.1 (the file of *tnmo-vnmo* pairs for CMP 60), par.2 (the file of *tnmo-vnmo* pairs for CMP 80), etc., into file *outpick*.

Line 445 uses SU command **mkparfile** to re-orient the columns of *par.cmp* to rows. File *par.cmp* holds the CMP numbers and their processing order:

```
60    1
80    2
100   3
etc.
```

Below is an example output file after processing two CMPs (line20: `numCMPs=2`).

```
cdp=60,80 \
#=1,2 \
tnmo=0.0158311,0.369393,0.69657,1.08707,1.98945 \
vnmo=1497,1526.7,2670.15,2952.3,3130.5 \
tnmo=0.0158311,0.353562,0.69657,1.05541,1.98945 \
vnmo=1497,1541.55,2655.3,2774.1,2878.05 \
```

Line 450 puts a blank space and a continuation mark, the backslash (\), at the end of each line (Lamb, 1990, pages 91-92).

7.6.6.9   iva.sh — Pause, Remove Temporary Files, Exit

Just before removing all temporary files and exiting, the name of the output file is written to the screen. We suggest that you rename the file to prevent writing over the picks with a later run.

7.6.6.10   iva.sh — vpick.txt

The seismic data input to interactive velocity analysis is *cmp4.su*. The output file of t-v picks is below, In Chapter 8, we call this file *vpick4.txt*.

```
 1    cdp=60,80,100,120,140,160,180,200,220,240,260,280,300,320,340,360,380,400 \
 2    #=1,2,3,4,5,6,7,8,9,10,11,12,13,14,15,16,17,18 \
 3    tnmo=0.0158311,0.390501,0.686016,1.06069,1.99472 \
 4    vnmo=1511.85,1511.85,2640.45,2922.6,3249.3 \
 5    tnmo=0.0158311,0.353562,0.701847,1.0343,1.99472 \
 6    vnmo=1482.15,1511.85,2625.6,2774.1,2982 \
 7    tnmo=0.0158311,0.364116,0.765172,0.875989,1.08707,1.98945 \
 8    vnmo=1511.85,1526.7,2595.9,2714.7,2625.6,2699.85 \
 9    tnmo=0.0211082,0.437995,0.844327,0.949868,1.16623,1.98945 \
10    vnmo=1497,1526.7,2551.35,2432.55,2685,2685 \
11    tnmo=0.0105541,0.569921,1.00792,1.30343,1.99472 \
12    vnmo=1497,1541.55,2595.9,2848.35,3011.7 \
13    tnmo=0.0158311,0.686016,1.12401,1.44591,1.98945 \
14    vnmo=1482.15,1541.55,2729.55,3160.2,3293.85 \
15    tnmo=0.0158311,0.78628,1.19789,1.5409,1.99472 \
16    vnmo=1482.15,1511.85,3011.7,3694.8,3798.75 \
17    tnmo=0.0105541,0.765172,1.1029,1.45119,1.98945 \
18    vnmo=1497,1511.85,2937.45,3665.1,3858.15 \
19    tnmo=0.0158311,0.627968,0.907652,1.25594,1.99472 \
20    vnmo=1497,1556.4,2595.9,3056.25,3353.25 \
21    tnmo=0.0158311,0.469657,0.686016,1.07124,1.98417 \
22    vnmo=1482.15,1556.4,2521.65,2818.65,3026.55 \
23    tnmo=0.0158311,0.3219,0.501319,0.923483,1.98417 \
24    vnmo=1482.15,1541.55,2462.25,2670.15,2774.1 \
25    tnmo=0.0158311,0.226913,0.395778,0.870712,1.98417 \
26    vnmo=1482.15,1511.85,2521.65,2699.85,2848.35 \
27    tnmo=0.0105541,0.23219,0.390501,0.918206,1.99472 \
28    vnmo=1482.15,1526.7,2566.2,2878.05,3234.45 \
29    tnmo=0.0158311,0.258575,0.46438,1.99472 \
30    vnmo=1497,1526.7,2699.85,3368.1 \
31    tnmo=0.0158311,0.311346,0.538259,0.612137,1.98945 \
32    vnmo=1482.15,1526.7,2759.25,2952.3,3442.35 \
33    tnmo=0.0211082,0.353562,0.622691,0.744063,1.99472 \
34    vnmo=1467.3,1511.85,2774.1,3056.25,3457.2 \
35    tnmo=0.0211082,0.369393,0.691293,0.812665,1.99472 \
36    vnmo=1482.15,1511.85,2878.05,2982,3279 \
37    tnmo=0.0211082,0.348285,0.728232,0.881266,1.20844,1.98945 \
38    vnmo=1482.15,1511.85,2833.5,3026.55,3145.35,3323.55 \
```

Note: Line 2 of this file must be removed before it can be used in a script. A comment line is not acceptable within a script command (see Section 8.2).

7.7   X-axis Labels in suxwigb and suximage

Program **suxwigb** allows you to supply a *key* value for labeling the x-axis. (In SU, the seismic x-axis is also called the slow dimension.) If you supply a *key* value; for example, *offset*, **suxwigb** correctly annotates offset values. It even plots the traces with irregular spacing if the *key* values do not occur at regular increments.

Program **suximage** accepts instructions for labeling the slow dimension (the seismic x-axis). Program **suximage** also has default instructions for labeling the x-axis; that is, a way to label the x-axis when the user does not supply an instruction.

Let's look at a small part of the **suximage** code (**suximage.c**, 2005):

```
153     if (!getparfloat("f2", &f2)) {
154             if      (tr.f2)         f2 = tr.f2;
155             else if (akey)          f2 = vtof(stype,val1) ;
156             else if (tr.tracr)      f2 = (float) tr.tracr;
157             else if (tr.tracl)      f2 = (float) tr.tracl;
158             else if (seismic)       f2 = 1.0;
159             else                    f2 = 0.0;
160     }
```

These lines are a hierarchy for creating the first value for the slow dimension.

- If the user supplied a value for parameter *f2*, use it.
- If the user supplied a value for *key*, use it.
- If trace header key *tracr* has a value, use it.
- If trace header key *tracl* has a value, use it.
- If this is seismic data, set the first value to "1."
- If none of the preceding is true, set the first value to "0."

In SU documentation, f2 is called "first trace location." By extension, all the parameters listed above only set the annotation value for the first displayed trace.

Now, let's look at trace (slow dimension) increment. Below is the line of code in **suximage.c** that refers to trace header key *d2*, "sample spacing between traces."

```
145     if (!getparfloat("d2", &d2)) d2 = (tr.d2) ? tr.d2 : 1.0;
```

This line creates a hierarchy for setting *d2*.

If the user supplied a value for *d2*, use it.

If trace header key *d2* has a value, use it.

If neither of the preceding is true, set *d2* to "1."

In conclusion:

- Program **suxwigb** reads each trace's *key* value and properly increments irregularly spaced values.

- Program **suximage**:
    - ○ Careful use of *f2* and *d2* can produce a result similar to the *key* parameter of **suxwigb**.
    - ○ The default **suximage** increment is "1,"so trace labeling after the first trace will be correct in only the simplest cases.

8.   Model 4: T-V Picks QC, NMO, Stack

8.1   Introduction

Now that we have a file of time-velocity (t-v) picks at selected CMPs, we check the quality of those picks. Then, we apply NMO to the data. After reviewing the results of NMO, it is simple (a one-line command) to stack the gathers.

For the sake of presentation in Section 8.2.1, what was one line in *vpick4.txt* (line 1) are now two lines in *tvQC.sh* (lines 13-14). Notice that *THERE ARE NO SPACES AT THE END OF LINE 13 AND NO SPACES AT THE BEGINNING OF LINE 14*.

We removed line 2 from file *vpick4.txt*. That line would be between lines 14 and 15 below. We removed it because *A COMMENT LINE WITHIN THE LINES OF AN SU COMMAND WILL MAKE THE SCRIPT CRASH*.

We also removed the continuation mark from the end of line 50.

8.2   Time-Velocity QC

At the end of interactive velocity analysis (Section 7.6.6.10), we have a file that contains a row of *cdp* values. Also within that file, there is a pair of *tnmo* values and *vnmo* values for each *cdp* value. Before we use these *cdp-tnmo-vnmo* values in normal moveout correction (NMO), we want to check the quality of our picks.

8.2.1   Program tvnmoqc, mode = 1

Each time series (set of *tnmo* values) input to the Seismic Un*x NMO program, **sunmo**, must increase in time. In other words, in a time series, each successive pick must be at a later time than the previous pick. The following script, *tvqc.sh*, offers a fast way to check that the *tnmo* values are acceptable to **sunmo**.

Script *tvQC.sh* uses program **tvnmoqc**; that program is presented in Appendix C. In *tvQC.sh*, we set mode=1. In this mode, **tvnmoqc** checks the number of *cdp* values, it checks that the number *tnmo-vnmo* pairs match the number of *cdp* values, and it checks that values in a *tnmo* series always increase. The output of **tvnmoqc** is screen messages.

```
 1  #! /bin/sh
 2  # File: tvQC.sh
 3  #          Test tnmo-vnmo series
 4  #   Input: The same cdp-tnmo-vnmo file that
 5  #              would be input to sunmo
 6  # Output: Screen messages
 7  # Use: tvQC.sh
 8
 9  # Set debugging on
10  ##set -x
11
12  tvnmoqc mode=1 \
13  cdp=60,80,100,120,140,160,180,200,220,\
14  240,260,280,300,320,340,360,380,400 \
15  tnmo=0.0158311,0.390501,0.686016,1.06069,1.99472 \
16  vnmo=1511.85,1511.85,2640.45,2922.6,3249.3 \
17  tnmo=0.0158311,0.353562,0.701847,1.0343,1.99472 \
18  vnmo=1482.15,1511.85,2625.6,2774.1,2982 \
19  tnmo=0.0158311,0.364116,0.765172,0.875989,1.08707,1.98945 \
20  vnmo=1511.85,1526.7,2595.9,2714.7,2625.6,2699.85 \
```

```
21    tnmo=0.0211082,0.437995,0.844327,0.949868,1.16623,1.98945 \
22    vnmo=1497,1526.7,2551.35,2432.55,2685,2685 \
23    tnmo=0.0105541,0.569921,1.00792,1.30343,1.99472 \
24    vnmo=1497,1541.55,2595.9,2848.35,3011.7 \
25    tnmo=0.0158311,0.686016,1.12401,1.04591,1.98945 \
26    vnmo=1482.15,1541.55,2729.55,3160.2,3293.85 \
27    tnmo=0.0158311,0.78628,1.19789,1.5409,1.99472 \
28    vnmo=1482.15,1511.85,3011.7,3694.8,3798.75 \
29    tnmo=0.0105541,0.765172,1.1029,1.45119,1.98945 \
30    vnmo=1497,1511.85,2937.45,3665.1,3858.15 \
31    tnmo=0.0158311,0.627968,0.907652,1.25594,1.99472 \
32    vnmo=1497,1556.4,2595.9,3056.25,3353.25 \
33    tnmo=0.0158311,0.469657,0.686016,1.07124,1.98417 \
34    vnmo=1482.15,1556.4,2521.65,2818.65,3026.55 \
35    tnmo=0.0158311,0.3219,0.501319,0.923483,1.98417 \
36    vnmo=1482.15,1541.55,2462.25,2670.15,2774.1 \
37    tnmo=0.0158311,0.226913,0.395778,0.870712,1.98417 \
38    vnmo=1482.15,1511.85,2521.65,2699.85,2848.35 \
39    tnmo=0.0105541,0.23219,0.390501,0.918206,1.99472 \
40    vnmo=1482.15,1526.7,2566.2,2878.05,3234.45 \
41    tnmo=0.0158311,0.258575,0.46438,1.99472 \
42    vnmo=1497,1526.7,2699.85,3368.1 \
43    tnmo=0.0158311,0.311346,0.538259,0.612137,1.98945 \
44    vnmo=1482.15,1526.7,2759.25,2952.3,3442.35 \
45    tnmo=0.0211082,0.353562,0.622691,0.744063,1.99472 \
46    vnmo=1467.3,1511.85,2774.1,3056.25,3457.2 \
47    tnmo=0.0211082,0.369393,0.691293,0.812665,1.99472 \
48    vnmo=1482.15,1511.85,2878.05,2982,3279 \
49    tnmo=0.0211082,0.348285,0.728232,0.881266,1.20844,1.98945 \
50    vnmo=1482.15,1511.85,2833.5,3026.55,3145.35,3323.55
51
52    # Exit politely from shell
53    exit
54
```

Before running *tvQC.sh*, we modified one time value on line 25. Below are the screen messages from *tvQC.sh*.

```
tvnmoqc: This file has 18 CDPs.

tvnmoqc: tnmo values must increase for use in NMO

tvnmoqc: For cdp=160, check times 1.12401 and 1.04591.

tvnmoqc: End of cdp-tnmo-vnmo check.
```

First, if there had been no time series problems, only the first and last lines would have printed to the screen. Second, no matter how many time series problems the input file has, **tvnmoqc** reads the entire file. It is up to you, the user, to correct time series problems before you use these lines for NMO (program **sunmo**). See Section 8.2.3.

8.2.2 Program tvnmoqc, mode = 2

Weeks or months after picking data, we might ask ourselves, "How good (accurate) were those picks?" The following interactive script, *velanQC*, answers that question. Script *velanQC* makes two images: the seismic CMP and the velan, and it puts the line of t-v picks on the velan.

Inside script velanQC, we use **tvnmoqc** with mode=2 and we supply a value for **tvnmoqc** parameter *prefix*. Program **tvnmoqc** takes each *tnmo-vnmo* pair, converts the rows of data to time-velocity (t-v) columns, and writes the columns to a file. The name of each file is "prefix value" "dot" "cdp value." For example, the first cdp in file *vpick4.txt* is 60. If we set `prefix=velqc`, the name of the first output file is *velqc.60*. The contents of *velqc.60* are:

```
0.0158311    1511.85
0.390501     1511.85
0.686016     2640.45
1.06069      2922.6
1.99472      3249.3
```

To overlay t-v picks on the velan, **suximage** needs data in columns, not rows.

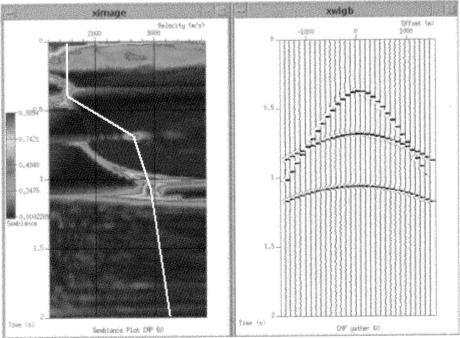

Figure 8.1: *Left: Velan of CMP 60. Right: CMP 60.*

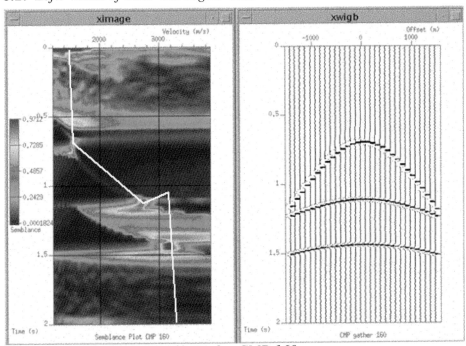

Figure 8.2: *Left: Velan of CMP 160. Right: CMP 160.*

Figure 8.1 shows CMP 60 with the velan and the picks on the velan. Figure 8.2 shows CMP 160 with the velan and the picks on the velan. The time value that we altered earlier has not been corrected, so we see the sharp up movement at the altered time. The contents of file *velqc.160* are:

```
0.0158311    1482.15
0.686016     1541.55
1.12401      2729.55
1.04591      3160.2
1.98945      3293.85
```

As you can see in the file at the end of Chapter 7, the fourth time, 1.04591 should be 1.44591.

Below are the first screen messages from *velanQC*. Lines numbers are added for discussion.

```
 1
 2   tvnmoqc: This file has 18 CDPs.
 3
 4   tvnmoqc: tnmo values must increase for use in NMO
 5
 6   tvnmoqc: For cdp=160, check times 1.12401 and 1.04591.
 7
 8   tvnmoqc: End of cdp-tnmo-vnmo check.
 9
10      *** VELOCITY ANALYSIS QC ***
11
12   Preparing CMP 1 of 18 for Display
13
14   Picks for CMP 60:
15   ------------------------
16   0.0158311    1511.85
17   0.390501     1511.85
18   0.686016     2640.45
19   1.06069      2922.6
20   1.99472      3249.3
21   ------------------------
22
23   Press Return or Enter to see next gather
24     Or enter "x" to exit
25   x
26   Terminated
27   kill: 6548: no such process
28   Terminated
29   kill: 6557: no such process
30
31        ==>  Closing velanQC
32
33   press return key to continue
34
```

Lines 1-8 are the same as are printed when mode=1 because the same checks are done to the *cdp-tnmo-vnmo* vales while **tvnmoqc** reads them. After error-checking and diagnostic printing, all the files of t-v values are created.

Lines 12-24 are printed as each t-v file is read from disk, and as the CMP and velan images are made on the screen.

In this example, instead of continuing to the next gather, the user enters "x" to exit the script: line 25. As a result, messages are printed to the screen as the seismic display windows are closed (lines 26-29) and the script announces that it is finished processing (line 31). After the user presses Enter or Return (as instructed on line 33), the script ends.

8.2.2.1 Script velanQC.scr

We usually start interactive scripts by running a *.scr* script. Here is *velanQC.scr*:

```
1   #! /bin/sh
2   # File: velanQC.scr
3   #       Run this script to start script velanQC.sh
4
5   xterm -geom 80x20+10+600 -e velanQC.sh
6
```

Line 1 invokes the shell. Line 5 opens a dialog window 80 characters wide by 20 characters high. Line 5 also places the dialog window 10 pixels from the left side of the viewing area and 600 pixels from the top of the viewing area. Line 5 also starts *velanQC.sh*, the processing script, within that window.

8.2.2.2 Script velanQC.sh

We start interactive scripts by running a *.scr* script. Here is *velanQC.sh*.

```
1   #! /bin/sh
2   # File: velanQC.sh
3   #       Run script velanQC.scr to start this script
4
5   # Set messages on
6   ##set -x
7
8   #=================================================
9   # USER AREA -- SUPPLY VALUES
10  #-------------------------------------------------
11  # CMPs for QC
12
13  cmp1=60   cmp2=80   cmp3=100
14  cmp4=120  cmp5=140  cmp6=160
15  cmp7=180  cmp8=200  cmp9=220
16  cmp10=240 cmp11=260 cmp12=280
17  cmp13=300 cmp14=320 cmp15=340
18  cmp16=360 cmp17=380 cmp18=400
19
20  numCMPs=18
21
22  #-------------------------------------------------
23  # Name of files
24  seisdata=../final/cmp4.su       # SU seismic
25  prefix=velqc          # tvnmoqc cdp-t-v files
26
27  #-------------------------------------------------
28  # display choices
29  myperc=98        # perc value for plot
30  plottype=0       # 0 = wiggle plot,  1 = image plot
31
32  #-------------------------------------------------
33  # Processing variables
34
35  # Semblance variables
36  nvs=100  # number of velocities
37  dvs=27   # velocity intervals
38  fvs=1200 # first velocity
39
40  #=================================================
```

```
41  # Insert pick file.
42  # Remove continuation mark (\) from last line.
43
44  tvnmoqc mode=2 prefix=$prefix \
45  cdp=60,80,100,120,140,160,180,200,220,\
46  240,260,280,300,320,340,360,380,400 \
47  tnmo=0.0158311,0.390501,0.686016,1.06069,1.99472 \
48  vnmo=1511.85,1511.85,2640.45,2922.6,3249.3 \
49  tnmo=0.0158311,0.353562,0.701847,1.0343,1.99472 \
50  vnmo=1482.15,1511.85,2625.6,2774.1,2982 \
51  tnmo=0.0158311,0.364116,0.765172,0.875989,1.08707,1.98945 \
52  vnmo=1511.85,1526.7,2595.9,2714.7,2625.6,2699.85 \
53  tnmo=0.0211082,0.437995,0.844327,0.949868,1.16623,1.98945 \
54  vnmo=1497,1526.7,2551.35,2432.55,2685,2685 \
55  tnmo=0.0105541,0.569921,1.00792,1.30343,1.99472 \
56  vnmo=1497,1541.55,2595.9,2848.35,3011.7 \
57  tnmo=0.0158311,0.686016,1.12401,1.04591,1.98945 \
58  vnmo=1482.15,1541.55,2729.55,3160.2,3293.85 \
59  tnmo=0.0158311,0.78628,1.19789,1.5409,1.99472 \
60  vnmo=1482.15,1511.85,3011.7,3694.8,3798.75 \
61  tnmo=0.0105541,0.765172,1.1029,1.45119,1.98945 \
62  vnmo=1497,1511.85,2937.45,3665.1,3858.15 \
63  tnmo=0.0158311,0.627968,0.907652,1.25594,1.99472 \
64  vnmo=1497,1556.4,2595.9,3056.25,3353.25 \
65  tnmo=0.0158311,0.469657,0.686016,1.07124,1.98417 \
66  vnmo=1482.15,1556.4,2521.65,2818.65,3026.55 \
67  tnmo=0.0158311,0.3219,0.501319,0.923483,1.98417 \
68  vnmo=1482.15,1541.55,2462.25,2670.15,2774.1 \
69  tnmo=0.0158311,0.226913,0.395778,0.870712,1.98417 \
70  vnmo=1482.15,1511.85,2521.65,2699.85,2848.35 \
71  tnmo=0.0105541,0.23219,0.390501,0.918206,1.99472 \
72  vnmo=1482.15,1526.7,2566.2,2878.05,3234.45 \
73  tnmo=0.0158311,0.258575,0.46438,1.99472 \
74  vnmo=1497,1526.7,2699.85,3368.1 \
75  tnmo=0.0158311,0.311346,0.538259,0.612137,1.98945 \
76  vnmo=1482.15,1526.7,2759.25,2952.3,3442.35 \
77  tnmo=0.0211082,0.353562,0.622691,0.744063,1.99472 \
78  vnmo=1467.3,1511.85,2774.1,3056.25,3457.2 \
79  tnmo=0.0211082,0.369393,0.691293,0.812665,1.99472 \
80  vnmo=1482.15,1511.85,2878.05,2982,3279 \
81  tnmo=0.0211082,0.348285,0.728232,0.881266,1.20844,1.98945 \
82  vnmo=1482.15,1511.85,2833.5,3026.55,3145.35,3323.55
83
84  #================================================
85
86  # HOW SEMBLANCE (VELAN) VELOCITIES ARE COMPUTED
87
88  # Last Vel =   fvs + (( nvs-1 ) * dvs ) = lvs
89  #     5000 =   500 + ((  99-1 ) * 45  )
90  #     3900 = 1200 + (( 100-1 ) * 27  )
91
92  # Compute last semblance (velan) velocity
93  lvs=`bc -l << -END
94  $fvs + (( $nvs - 1 ) * $dvs )
95  END`
96
97  #================================================
98
99  # FILE DESCRIPTIONS
100
101 # tmp1 = ASCII temp file to avoid screen display of "zap"
102 # tmp2 = ASCII temp file for "wc" result (velan)
103 # tmp3 = ASCII temp file for stripping file name from tmp2 (velan)
```

```
104  # panel.$picknow = current CMP windowed from line of CMPs (seismic)
105  # ${prefix}.${picknow} = current CMP picks arranged as "t1 v1"
106  #                                                           "t2 v2"
107  #                                                           etc.
108
109  #=================================================
110
111  echo " "
112  echo "   *** VELOCITY ANALYSIS QC ***"
113  echo " "
114
115  #-------------------------------------------------
116
117  # Remove old files.
118  rm -f panel.* picks.* tmp*
119
120  #-------------------------------------------------
121  # Get ns, dt, first time from seismic file
122  nt=`sugethw ns < $seisdata | sed 1q | sed 's/.*ns=//'`
123  dt=`sugethw dt < $seisdata | sed 1q | sed 's/.*dt=//'`
124  ft=`sugethw delrt < $seisdata | sed 1q | sed 's/.*delrt=//'`
125
126  # Convert dt from header value in microseconds
127  # to seconds for velocity profile plot
128  dt=`bc -l << -END
129    scale=6
130    $dt / 1000000
131  END`
132
133  # If "delrt", use it; else use zero
134  if [ $ft -ne 0 ] ; then
135    tstart=`bc -l << -END
136      scale=6
137      $ft / 1000
138    END`
139  else
140    tstart=0.0
141  fi
142
143  #=================================================
144  # BEGIN QC LOOP
145  #-------------------------------------------------
146
147  i=1
148  while [ $i -le $numCMPs ]
149  do
150
151  # set variable $picknow to current CMP
152    eval picknow=\$cmp$i
153
154    echo " "
155    echo "Preparing CMP $i of $numCMPs for Display "
156    echo " "
157
158  #-------------------------------------------------
159  # Plot CMP (right)
160  #-------------------------------------------------
161
162    suwind < $seisdata \
163            key=cdp min=$picknow max=$picknow > panel.$picknow
164
165    if [ $plottype -eq 0 ] ; then
166      suxwigb < panel.$picknow xbox=422 ybox=10 wbox=400 hbox=550 \
```

```
167                    title="CMP gather $picknow" \
168                    label1=" Time (s)" label2="Offset (m)" key=offset \
169                    perc=$myperc verbose=0 &
170      else
171        suximage < panel.$picknow xbox=421 ybox=10 wbox=400 hbox=550 \
172                    title="CMP gather $picknow" \
173                    label1=" Time (s)" \
174                    perc=$myperc verbose=0 &
175      fi
176
177    #-------------------------------------------------
178    # Plot semblance (velan) (left)
179    #-------------------------------------------------
180
181    # Get the number of picks (number of lines) in prefix.picknow |
182    #   Remove blank spaces preceding the line count.
183    # Remove file name that was returned from "wc".
184    # Store line count in "npair" to guide line on velan.
185
186      echo "Picks for CMP ${picknow}:"
187      echo "------------------------"
188      cat ${prefix}.${picknow}
189      echo "------------------------"
190
191      wc -l ${prefix}.${picknow} | sed 's/^  *\(.*\)/\1/' > tmp2
192      sed 's/  ${prefix}.${picknow}  //' < tmp2 > tmp3
193      npair=`sort < tmp3`
194
195      suvelan < panel.$picknow nv=$nvs dv=$dvs fv=$fvs |
196      suximage xbox=10 ybox=10 wbox=400 hbox=550 perc=99 \
197                units="semblance" f2=$fvs d2=$dvs n2tic=5 \
198                title="Semblance Plot CMP $picknow" cmap=hsv2 \
199                label1=" Time (s)" label2="Velocity (m/s)" \
200                legend=1 units=Semblance verbose=0 gridcolor=black \
201                grid1=solid grid2=solid \
202                curve=${prefix}.${picknow} npair=$npair curvecolor=white &
203
204    #-------------------------------------------------
205    # Question: Next CMP/Velan or Exit?
206    #-------------------------------------------------
207
208      echo " "
209      echo "Press Return or Enter to see next gather"
210      echo "  Or enter \"x\" to exit"
211      > /dev/tty
212      read response
213
214      case $response in
215        [xX])
216              zap xwigb > tmp1
217              zap ximage > tmp1
218              i=`expr $numCMPs + 1`
219              ;;
220          *)
221              zap xwigb > tmp1
222              zap ximage > tmp1
223              i=`expr $i + 1`
224              ;;
225      esac
226
227  done
228
229    #=================================================
```

```
230   # Exit
231   #---------------------------------------------
232
233   # Exit
234   echo " "
235   echo "      ==>  Closing velanQC"
236   pause
237
238   rm -f panel.* picks.* tmp*
239   ##rm -f $prefix.*
240   exit
241
```

Script *velanQC.sh* is simpler than *iva.sh*; it only makes two displays and it continuously moves forward through the gathers. You only choose whether to view the next gather and velan with its picks or to quit the script.

Program **tvnmoqc** is called on line 44. The pick file must be placed immediately after.

Note: Remove the last continuation mark (\) from the pick file once it is inside *velanQC.sh*. Set the value for **tvnmoqc** parameter *prefix* on line 25.

Unlike *iva.sh* user parameters, here the user does not supply an *outpick* file name and does not supply CVS variables.

The CMPs for QC (lines 13-18) have to be the same CMPs listed in *vpick4.txt*. You can omit CMPs, but the ones included must be the same, or the picks will not properly overlay.

Note: When *velanQC.sh* finishes, it does not remove the individual t-v files. However, if you un-comment line 239 (remove "##"), it will remove the files.

8.2.3 Edit Time Picks

What do we do when we find a series with time values that do not increase? Here, we provide three answers. Before we look at the answers, let's restate the problem.

Pair 1 shows a *tnmo-vnmo* series with times and velocities that always increase. We created this pair in Chapter 7.

```
Pair 1:
------
 25   tnmo=0.0158311,0.686016,1.12401,1.44591,1.98945 \
 26   vnmo=1482.15,1541.55,2729.55,3160.2,3293.85 \
```

Pair 2 shows the *tnmo-vnmo* series that we used in our examples above. A mistaken pick like this is possible, but not common. We do not usually move the mouse up as we make picks down the velan.

```
Pair 2:
------
 25   tnmo=0.0158311,0.686016,1.12401,1.04591,1.98945 \
 26   vnmo=1482.15,1541.55,2729.55,3160.2,3293.85 \
```

Pair 3 shows a more common picking mistake. Here we made two picks on top of each other. The third and fourth picks are the same – we did not move the mouse between picks. The result is an extra pick between two intended picks.

```
Pair 3:
------
 25   tnmo=0.0158311,0.686016,1.12401,1.12401,1.44591,1.98945 \
 26   vnmo=1482.15,1541.55,2729.55,2729.55,3160.2,3293.85 \
```

Let's solve the problem presented by Pair 3.

Our first solution is to remove the extra time-velocity pair. Once we do that, we have the same *tnmo-vnmo* series as Pair 2. This is a good solution, but we try not to use it. We are afraid of making a mistake while typing and backspacing in this crowded file.

```
Solution 1:
----------
25   tnmo=0.0158311,0.686016,1.12401,1.44591,1.98945 \
26   vnmo=1482.15,1541.55,2729.55,3160.2,3293.85 \
```

Our second solution is to increase the value of the fourth time by one millisecond (we do not change the corresponding velocity). We change the fourth time from 1.12401 to 1.12501. This change has just about no effect on NMO, and we spend less time editing solution 2 than we do solution 1.

```
Solution 2:
----------
25   tnmo=0.0158311,0.686016,1.12401,1.12501,1.44591,1.98945 \
26   vnmo=1482.15,1541.55,2729.55,2729.55,3160.2,3293.85 \
```

Our third solution is to re-pick the CMP. Below are lines from *iva.sh* (Section 7.6.6.3).

```
10   #------------------------------------------------
11   # CMPs for analysis
12
13   cmp1=60    cmp2=80    cmp3=100
14   cmp4=120   cmp5=140   cmp6=160
15   cmp7=180   cmp8=200   cmp9=220
16   cmp10=240  cmp11=260  cmp12=280
17   cmp13=300  cmp14=320  cmp15=340
18   cmp16=360  cmp17=380  cmp18=400
19
20   numCMPs=18
21
22   #------------------------------------------------
```

To re-pick CMP 160, we change the lines as shown below. What was "cmp6" is now "cmp1," and parameter *numCMPs* is now "1." Nothing else in *iva.sh* needs to change.

```
10   #------------------------------------------------
11   # CMPs for analysis
12
13
14   cmp1=160
15
16
17
18
19
20   numCMPs=1
21
22   #------------------------------------------------
```

After we re-pick CMP 160, we replace the old *tnmo-vnmo* pair with the new *tnmo-vnmo* pair.

8.3   *Normal Moveout: Script nmo4.sh*

The file *vpick4.txt* that was shown at the end of Chapter 7 is now presented within script *nmo4.sh*. As we noted in Section 8.2.1, what was one line in *vpick4.txt* (line 1) is now two lines in *nmo4.sh* (lines 18-19). Also, we removed line 2 from file *vpick4.txt* (refer to the end of Chapter 7). That line would be between lines 19 and 20 below.

```
 1  #! /bin/sh
 2  # File: nmo4.sh
 3  #        Apply NMO (flatten) 2-D line of CMPs
 4  # Input (1): 2-D line of CMPs
 5  # Output (1): NMO-corrected 2-D line of CMPs
 6  # Use: nmo4.sh
 7  #
 8  # NMO correction is interpolated between named CMPs.
 9
10  # Set debugging on
11  set -x
12
13  # Name data sets
14  indata=cmp4.su
15  outdata=nmo4.su
16
17  sunmo < $indata \
18  cdp=60,80,100,120,140,160,180,200,220,\
19  240,260,280,300,320,340,360,380,400 \
20  tnmo=0.0158311,0.390501,0.686016,1.06069,1.99472 \
21  vnmo=1511.85,1511.85,2640.45,2922.6,3249.3 \
22  tnmo=0.0158311,0.353562,0.701847,1.0343,1.99472 \
23  vnmo=1482.15,1511.85,2625.6,2774.1,2982 \
24  tnmo=0.0158311,0.364116,0.765172,0.875989,1.08707,1.98945 \
25  vnmo=1511.85,1526.7,2595.9,2714.7,2625.6,2699.85 \
26  tnmo=0.0211082,0.437995,0.844327,0.949868,1.16623,1.98945 \
27  vnmo=1497,1526.7,2551.35,2432.55,2685,2685 \
28  tnmo=0.0105541,0.569921,1.00792,1.30343,1.99472 \
29  vnmo=1497,1541.55,2595.9,2848.35,3011.7 \
30  tnmo=0.0158311,0.686016,1.12401,1.44591,1.98945 \
31  vnmo=1482.15,1541.55,2729.55,3160.2,3293.85 \
32  tnmo=0.0158311,0.78628,1.19789,1.5409,1.99472 \
33  vnmo=1482.15,1511.85,3011.7,3694.8,3798.75 \
34  tnmo=0.0105541,0.765172,1.1029,1.45119,1.98945 \
35  vnmo=1497,1511.85,2937.45,3665.1,3858.15 \
36  tnmo=0.0158311,0.627968,0.907652,1.25594,1.99472 \
37  vnmo=1497,1556.4,2595.9,3056.25,3353.25 \
38  tnmo=0.0158311,0.469657,0.686016,1.07124,1.98417 \
39  vnmo=1482.15,1556.4,2521.65,2818.65,3026.55 \
40  tnmo=0.0158311,0.3219,0.501319,0.923483,1.98417 \
41  vnmo=1482.15,1541.55,2462.25,2670.15,2774.1 \
42  tnmo=0.0158311,0.226913,0.395778,0.870712,1.98417 \
43  vnmo=1482.15,1511.85,2521.65,2699.85,2848.35 \
44  tnmo=0.0105541,0.23219,0.390501,0.918206,1.99472 \
45  vnmo=1482.15,1526.7,2566.2,2878.05,3234.45 \
46  tnmo=0.0158311,0.258575,0.46438,1.99472 \
47  vnmo=1497,1526.7,2699.85,3368.1 \
48  tnmo=0.0158311,0.311346,0.538259,0.612137,1.98945 \
49  vnmo=1482.15,1526.7,2759.25,2952.3,3442.35 \
50  tnmo=0.0211082,0.353562,0.622691,0.744063,1.99472 \
51  vnmo=1467.3,1511.85,2774.1,3056.25,3457.2 \
52  tnmo=0.0211082,0.369393,0.691293,0.812665,1.99472 \
53  vnmo=1482.15,1511.85,2878.05,2982,3279 \
54  tnmo=0.0211082,0.348285,0.728232,0.881266,1.20844,1.98945 \
55  vnmo=1482.15,1511.85,2833.5,3026.55,3145.35,3323.55 \
56         > $outdata
57
58  # Exit politely from shell
59  exit
60
```

This script runs very quickly.

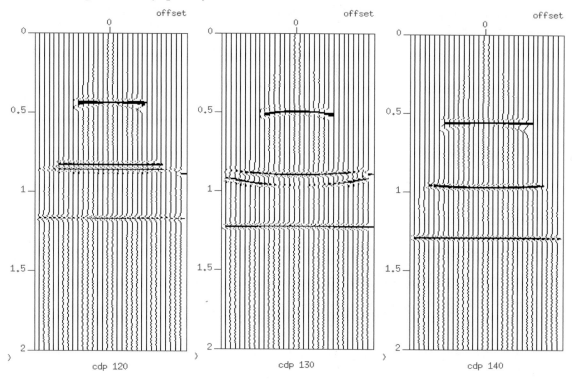

Figure 8.3: *Left: NMO applied to CMP 120. Center: NMO applied to CMP 130. Right: NMO applied to CMP 140. For CMP 130, the NMO t-v values were linearly interpolated between CMP 120 and CMP 140.*

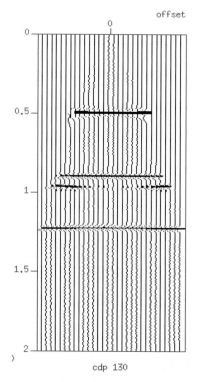

Figure 8.4: *NMO applied to CMP 130, no interpolation.*

Recall that t-v picks were made every 20th CMP starting at the first full-fold CMP, CMP 60, and ending at the last full-fold CMP, CMP 400 (see Section 7.5). Program **sunmo** used the values of script *nmo4.sh* to interpolate between CMPs.

How accurate is the interpolation? Mathematically, it is quite accurate. How useful is the interpolation? Figure 8.3 shows CMPs 120, 130 and 140 after NMO. We think we made precise picks to flatten CMPs 120 and 140; their reflectors look reasonably flat. But the third reflector of CMP 130, with interpolated picks, appears to curve quite a bit. When we stack CMP 130, we expect that the third reflector will smear; that is, it will not stack to a zero-offset spike. What is worse, the smear might contribute to smearing the second reflector. (We are showing the worst CMP of *nmo4.su*.)

To improve our stack, we modify script *iva.sh* to pick only CMP 130. After doing so, we show the flattened gather in Figure 8.4.

This is not perfect flattening, but better than before. Below is the new NMO script, nmo4a.sh.

```
 1   #! /bin/sh
 2   # File: nmo4a.sh
 3   #       Apply NMO (flatten) 2-D line of CMPs
 4   # Input (1): 2-D line of CMPs
 5   # Output (1): NMO-corrected 2-D line of CMPs
 6   # Use: nmo4a.sh
 7   #
 8   # NMO correction is interpolated between named CMPs.
 9
10   # Set debugging on
11   set -x
12
13   # Name data sets
14   indata=cmp4.su
15   outdata=nmo4a.su
16
17   sunmo < $indata \
18   cdp=60,80,100,120,130,140,160,180,200,220,\
19   240,260,280,300,320,340,360,380,400 \
20   tnmo=0.0158311,0.390501,0.686016,1.06069,1.99472 \
21   vnmo=1511.85,1511.85,2640.45,2922.6,3249.3 \
22   tnmo=0.0158311,0.353562,0.701847,1.0343,1.99472 \
23   vnmo=1482.15,1511.85,2625.6,2774.1,2982 \
24   tnmo=0.0158311,0.364116,0.765172,0.875989,1.08707,1.98945 \
25   vnmo=1511.85,1526.7,2595.9,2714.7,2625.6,2699.85 \
26   tnmo=0.0211082,0.437995,0.844327,0.949868,1.16623,1.98945 \
27   vnmo=1497,1526.7,2551.35,2432.55,2685,2685 \
28   tnmo=0.0211082,0.506596,0.912929,1.05013,1.24011,1.99472 \
29   vnmo=1511.85,1526.7,2595.9,2982,2714.7,3145.35 \
30   tnmo=0.0105541,0.569921,1.00792,1.30343,1.99472 \
31   vnmo=1497,1541.55,2595.9,2848.35,3011.7 \
32   tnmo=0.0158311,0.686016,1.12401,1.44591,1.98945 \
33   vnmo=1482.15,1541.55,2729.55,3160.2,3293.85 \
34   tnmo=0.0158311,0.78628,1.19789,1.5409,1.99472 \
35   vnmo=1482.15,1511.85,3011.7,3694.8,3798.75 \
36   tnmo=0.0105541,0.765172,1.1029,1.45119,1.98945 \
37   vnmo=1497,1511.85,2937.45,3665.1,3858.15 \
38   tnmo=0.0158311,0.627968,0.907652,1.25594,1.99472 \
39   vnmo=1497,1556.4,2595.9,3056.25,3353.25 \
40   tnmo=0.0158311,0.469657,0.686016,1.07124,1.98417 \
41   vnmo=1482.15,1556.4,2521.65,2818.65,3026.55 \
42   tnmo=0.0158311,0.3219,0.501319,0.923483,1.98417 \
43   vnmo=1482.15,1541.55,2462.25,2670.15,2774.1 \
```

```
44  tnmo=0.0158311,0.226913,0.395778,0.870712,1.98417 \
45  vnmo=1482.15,1511.85,2521.65,2699.85,2848.35 \
46  tnmo=0.0105541,0.23219,0.390501,0.918206,1.99472 \
47  vnmo=1482.15,1526.7,2566.2,2878.05,3234.45 \
48  tnmo=0.0158311,0.258575,0.46438,1.99472 \
49  vnmo=1497,1526.7,2699.85,3368.1 \
50  tnmo=0.0158311,0.311346,0.538259,0.612137,1.98945 \
51  vnmo=1482.15,1526.7,2759.25,2952.3,3442.35 \
52  tnmo=0.0211082,0.353562,0.622691,0.744063,1.99472 \
53  vnmo=1467.3,1511.85,2774.1,3056.25,3457.2 \
54  tnmo=0.0211082,0.369393,0.691293,0.812665,1.99472 \
55  vnmo=1482.15,1511.85,2878.05,2982,3279 \
56  tnmo=0.0211082,0.348285,0.728232,0.881266,1.20844,1.98945 \
57  vnmo=1482.15,1511.85,2833.5,3026.55,3145.35,3323.55 \
58        > $outdata
59
60  # Exit politely from shell
61  exit
62
```

Script *nmo4a*.sh is two lines longer than *nmo4.sh*. Line 18 is modified to include "130." The t-v picks for CMP 130 are the fifth *tnmo-vnmo* pair, lines 28-29.

8.4   Interactive Viewer

We can quickly review the gathers because we have an interactive viewer. Actually, like *velanQC*, the only interactivity that scripts *iview.scr* and *iview.sh* offer is when the user chooses to see the next gather or quit the script.

8.4.1 Script iview.scr

We usually start interactive scripts by running a *.scr* script. Here is *iview.scr*:

```
1  #! /bin/sh
2  # File: iview.scr
3  #         Run this script to start script iview.sh
4
5  xterm -geom 80x20+10+545 -e iview.sh
6
```

Line 1 invokes the shell. Line 5 opens a dialog window 80 characters wide by 20 characters high. Line 5 also places the dialog window 10 pixels from the left side of the viewing area and 545 pixels from the top of the viewing area. Line 5 also starts *iview.sh*, the processing script, within that window.

8.4.2 Script iview.sh

We start interactive scripts by running a *.scr* script. Here is *iview.sh*.

```
1  #! /bin/sh
2  # File: iview.sh
3  #         View seismic gathers from a 2-D line
4
5  # Set messages on
6  ##set -x
7
8  #===============================================
9  # USER AREA -- SUPPLY VALUES
10
11  # input seismic data
12  indata=nmo4a.su  # SU format
13
14  # plot choices
```

```
15   myperc=98          # perc value for plot
16   plottype=0         # 0 = wiggle plot,  1 = image plot
17   Wplot=400          # Width of plot (pixels)
18   Hplot=500          # Height of plot (pixels)
19
20   # processing variables
21   sortkey=cdp        # sort key (usually fldr or cdp)
22   firsts=10          # first sort (fldr or cdp) value
23    lasts=458         # last sort (fldr or cdp) value
24   increment=10       # sort key increment
25   tracekey=offset    # trace label key
26
27   #================================================
28   # file descriptions
29
30   # tmp1 = binary temp file for input gather
31
32   #------------------------------------------------
33
34   echo " "
35   echo "  *** VIEWER  ***"
36   echo " "
37   echo "      INPUT: $indata"
38
39   #------------------------------------------------
40   # Remove old temporary file
41   rm -f tmp*
42
43   #------------------------------------------------
44   # BEGIN LOOP
45   #------------------------------------------------
46
47   i=$firsts
48   while [ $i -le $lasts ]
49   do
50
51     echo " "
52     echo "Reading gather $i of $indata"
53     echo "First gather = $firsts    Last gather = $lasts"
54
55     suwind < $indata key=$sortkey min=$i max=$i > tmp1
56
57     if [ $plottype -eq 0 ] ; then
58       suxwigb < tmp1 xbox=10 ybox=10 wbox=$Wplot hbox=$Hplot \
59               title="$sortkey $i" \
60               label1=" Time (s)" label2="$tracekey" key=$tracekey \
61               perc=$myperc verbose=0 &
62     else
63       suximage < tmp1 xbox=10 ybox=10 wbox=$Wplot hbox=$Hplot \
64                title="$sortkey $i" \
65                label1=" Time (s)" \
66                perc=$myperc verbose=0 &
67     fi
68
69     echo " "
70     echo "Press Return or Enter to see next gather"
71     echo "  Or enter \"x\" to exit"
72     > /dev/tty
73     read response
74
75     case $response in
76       [xX])
77             zap xwigb > tmp1
```

```
 78              zap ximage > tmp1
 79              i=`expr $lasts + 1`
 80            ;;
 81          *)
 82              zap xwigb > tmp1
 83              zap ximage > tmp1
 84              i=`expr $i + $increment`
 85            ;;
 86     esac
 87
 88  done
 89
 90  #------------------------------------------------
 91  # END LOOP
 92  #------------------------------------------------
 93
 94  # Exit
 95  echo " "
 96  echo "       ==>   Closing iview"
 97  pause
 98  rm -f tmp*
 99  exit
100
```

The user-supplied values are on lines 8-26.

The gather loop is lines 43-92. Line 48 tests whether the value of "i" is within the processing range. If we decide to exit the script before viewing all gathers (by entering "x"), line 79 sets the value of "i" to one larger than variable *lasts* to make the script jump out of the view loop.

There are only two steps to using the script:

1. Click in the dialog window to make it active.

2. Press Return or Enter to see the next gather or enter "X" to exit the viewer.

Following is the screen dialogue of the first two gathers and the last two gathers (excluding system messages):

```
  *** VIEWER   ***

    INPUT: nmo4a.su

Reading gather 10 of nmo4a.su
First gather = 10     Last gather = 458

Press Return or Enter to see next gather
  Or enter "x" to exit

Reading gather 20 of nmo4a.su
First gather = 10     Last gather = 458

Press Return or Enter to see next gather
  Or enter "x" to exit


 . . .

Reading gather 440 of nmo4a.su
First gather = 10     Last gather = 458

Press Return or Enter to see next gather
  Or enter "x" to exit
```

```
Reading gather 450 of nmo4a.su
First gather = 10    Last gather = 458

Press Return or Enter to see next gather
  Or enter "x" to exit

     ==>  Closing iview

press return key to continue
```

8.5   Stack

A one-line command is sufficient to stack the data. The default stack key is *cdp*.

sustack < nmo4a.su > stack4.su

Below is the **surange** output of file *stack4.su*.

surange < stack4.su

```
458 traces:
 tracl=(1,458)  tracr=(1,12000)  fldr=(1,200)  tracf=(1,60)  cdp=(1,458)
 trid=1 nhs=(1,30)  sx=(0,9950)  gx=(-1475,11425)  ns=501
 dt=4000
```

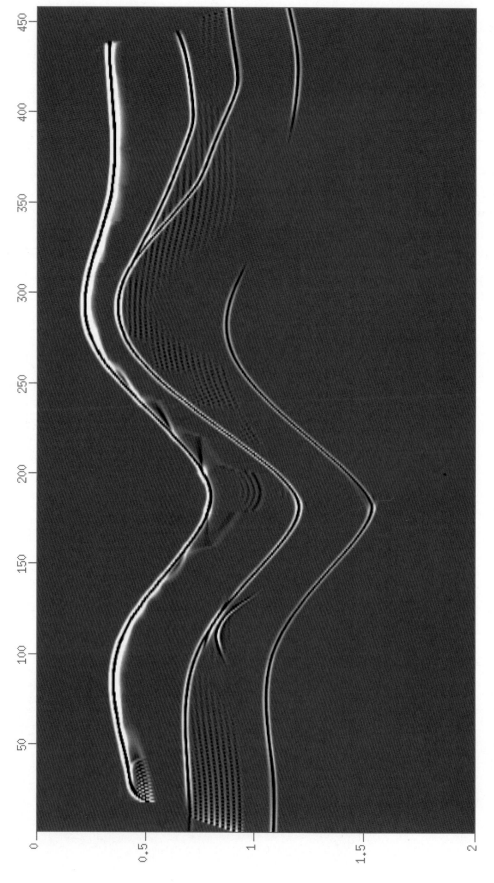

Figure 8.5: *An **ximage** plot of stack4.su. The horizontal axis is CMP number. The sand channel diffraction is near CMP 110, at approximately 0.8 seconds.*

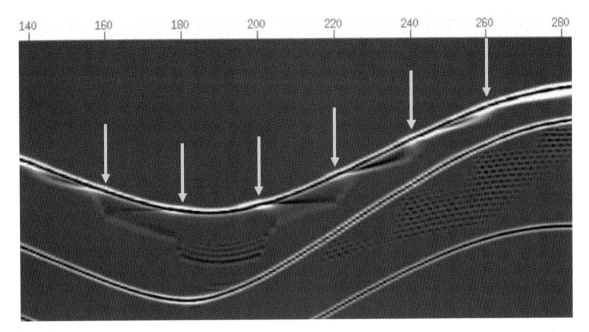

Figure 8.6: *Arrows point to velocity analysis CMP locations. Between those CMPs, the top reflector thickens.*

When we look at steep portions of the top reflector (Figure 8.6), we see that at CMP locations where we did velocity analysis, the top reflector is thin. We might say it is well focused. Between velocity analysis CMPs where **sunmo** interpolated t-v values, the top reflector is broader (out of focus, diffuse). The lesson: The steeper the reflector, the more often velocity analysis should be done.

Another lesson (Figure 8.7): Velocity analysis should be done more frequently where velocity changes rapidly (vertically or laterally). The channel on the left side of the stack, the diffraction with an apex near CMP 110 at approximately 0.8 seconds, is such a velocity discontinuity.

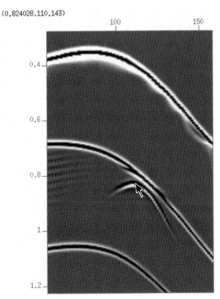

Figure 8.7: *The arrow points to a diffraction hyperbola, the time section expression of the sand channel.*

The *sfill* values we used to build Model 4 (Section 6.2) are the rms velocities. Since we know the rms velocities, we will migrate a portion of Model 4, the sand channel diffractor, to determine the correct value of **sustolt** parameter *vscale*. Once we know the value of *vscale* that collapses the sand channel diffraction using a velocity of approximately 3000 m/s, we will migrate the entire model.

9.2.3 Migration Constant Velocity Panels (CVPs)

We use migration program **sustolt** because it is fast. Here, we use it to create a series of constant velocity migrated sections. Script *migcvp.sh* takes in one stack section, migrates it several times, each time with a constant velocity. We call each migration product a "panel." All panels are together in the same output file (Figure 9.2). After migration, we use *iviewcvp.scr* and *iviewcvp.sh* (Section 9.3), variations of *iview.scr* and *iview.sh* (Section 8.4) to review the migration panels.

When we used **surange** on *stack4.su* (Section 8.5), we learned that the range of *cdp* values is 1-458. We can supply these values in *migcvp.sh* for variables *cdpmin* and *cdpmax* (lines 20-21). However, let us first examine the sand channel diffraction near CMP 110. The sand channel is a small event for which we know the model velocity: 3015 m/s (Table 6.2, Section 6.2). We can use the model velocity to determine **sustolt** parameter *vscale*.

```
                        +------------------+
                        |      Input       |
                        |     Seismic      |
                        |      Data        |
                        +------------------+
+-----------+-----------+-----------+-----------+-----------+-----------+
| Output of | Output of | Output of | Output of | Output of | Output of |
| Migration | Migration | Migration | Migration | Migration | Migration |
| Velocity 1| Velocity 2| Velocity 3| Velocity 4| Velocity 5| Velocity 6|
+-----------+-----------+-----------+-----------+-----------+-----------+
```

Figure 9.2: *When the input to script migcvp.sh contains 50 traces and migcvp.sh uses six different velocities, the output file contains 300 traces.*

We use lines 20 and 21 to select a window of CMPs around CMP 110; we migrate CMPs 80-150. Line 22 uses our calculated value of CDP bin spacing.

Since *migcvp.sh* does constant velocity migration

- we supplied 1.0 for *smig*, the stretch factor (line 23).
- we supplied "0" for *tmig* (line 79).

What should we use for the value of *vscale*? After several tests, we found that a velocity scale factor of 1.9 (line 25) yielded the best migration of the sand channel. We determined *vscale* by repeatedly migrating the sand channel diffraction until the best migration image was obtained for a migration velocity of 3000 m/s, the value we know to be the true interval velocity. Figure 9.3 was obtained using *vscale=1.5* and Figure 9.4 was obtained using *vscale=1.9*.

(In seismic data processing, we sometimes have to know the answer to get the right answer!)

As discussed in the previous section, the length of the taper on each side (*lstaper*) is set to 20 traces (line 26) and the length of the bottom taper (*lbtaper*) is set to 100 samples (line 27).

Lines 30-32 specify that the first panel is migrated at 1200 m/s, the last panel is migrated at 5000 m/s, and the panel interval is 200 m/s. These numbers mean *migcvp.sh* migrates the input data 20 times (see line 41):

$$((5000-1200) / 200) + 1 = 20$$

Line 34 is for testing. Set variable *numVtest* to "1" or "2" or another small value to limit the number of panels processed (lines 45-47).

```
1   #! /bin/sh
2   # File: migcvp.sh
3   #       Create one panel for each migration velocity
4   #       Each panel has the same "fldr" value
5   #       The migration velocity is in key "offset"
6   #       Total number of panels is in key "nvs"
7
8   # Set messages on
9   ##set -x
10
11  #================================================
12  # USER AREA -- SUPPLY VALUES
13  #------------------------------------------------
14
15  # Seismic files
16  indata=stack4.su     # SU format
17  outdata=migcvp.su    # migration Constant Velocity Panels
18
19  # Migration variables
20  cdpmin=80            # Start CDP value
21  cdpmax=150           # End CDP value
22  dxcdp=70.71          # distance between adjacent CDP bins (m)
23  smig=1.0             # stretch factor (0.6 typical if vrms increasing)
24                       # [the "W" factor] (Default=1.0)
25  vscale=1.9           # scale factor to apply to velocities (Default=1.0)
26  lstaper=20           # length of side tapers (traces) (Default=0)
27  lbtaper=100          # length of bottom taper (samples) (Default=0)
28
29  # Velocity panel variables
30  firstv=1200          # first velocity value
31   lastv=5000          # last velocity value
32  increment=200        # velocity increment
33
34  numVtest=100         # use to limit number of velocity panels
35                       # otherwise, use very large value (100)
36
37  #================================================
38
39  # Compute number of velocity panels
40
41  numV=`bc -l << -END
42  ( ( $lastv - $firstv ) / $increment ) + 1
43  END`
44
45  if [ $numVtest -lt $numV ] ; then
46    numV=$numVtest
47  fi
48
49  #------------------------------------------------
50
51  # FILE DESCRIPTIONS
52  # tmp1 = binary temp file of input data
53
54  #------------------------------------------------
```

```
55
56  cp $indata tmp1
57  migV=$firstv
58  echo " "
59
60  #------------------------------------------------
61  # Loop through Migration Constant Velocity Panels
62  #      Each panel has the same "fldr" value
63  #      Panel migration velocity is in key "offset"
64  #      Total number of panels (numV) is in key "nvs"
65  #------------------------------------------------
66
67  i=1
68  while [ $i -le $numV ]
69  do
70
71    echo "    iteration number = $i    Velocity = $migV"
72
73    suwind < tmp1 key=cdp min=$cdpmin max=$cdpmax |
74          sushw key=fldr a=$i |
75          sushw key=offset a=$migV |
76          sushw key=nvs a=$numV |
77
78    sustolt cdpmin=$cdpmin cdpmax=$cdpmax dxcdp=$dxcdp \
79          tmig=0 vmig=$migV smig=$smig vscale=$vscale \
80          lstaper=$lstaper lbtaper=$lbtaper \
81        >> $outdata
82
83    i=`expr $i + 1`
84    migV=`expr $migV + $increment`
85
86  done
87
88  #------------------------------------------------
89  # Remove files and exit
90  #------------------------------------------------
91
92  echo " "
93  echo "    Output file = $outdata"
94  echo " "
95
96  rm -f tmp*
97  exit
98
```

Line 56 copies the input data to a temporary file (*tmp1*) to avoid changing the original data. Line 57 sets the first migration velocity.

The loop is lines 60-87. Line 71 writes a message to the screen: the current migration panel number and its migration velocity. Line 73 windows the appropriate CMPs from the temporary file. Line 74 sets key *fldr* to the iteration counter *i* for all traces in a panel. Line 75 puts the migration velocity in key *offset* for all traces in a panel. Line 76 puts the number of panels in key *nvs* of every trace. Stolt migration is done in lines 78-80. Line 81 concatenates the output panels. Line 83 increments the loop (and panel) counter. Line 84 increments the velocity for the next panel.

Below are the screen messages for this run:

```
iteration number = 1    Velocity = 1200
iteration number = 2    Velocity = 1400
iteration number = 3    Velocity = 1600
iteration number = 4    Velocity = 1800
```

```
iteration number = 5      Velocity = 2000
iteration number = 6      Velocity = 2200
iteration number = 7      Velocity = 2400
iteration number = 8      Velocity = 2600
iteration number = 9      Velocity = 2800
iteration number = 10     Velocity = 3000
iteration number = 11     Velocity = 3200
iteration number = 12     Velocity = 3400
iteration number = 13     Velocity = 3600
iteration number = 14     Velocity = 3800
iteration number = 15     Velocity = 4000
iteration number = 16     Velocity = 4200
iteration number = 17     Velocity = 4400
iteration number = 18     Velocity = 4600
iteration number = 19     Velocity = 4800
iteration number = 20     Velocity = 5000

Output file = migcvp.su
```

9.3   Interactive Constant Velocity Panel Viewer

To view the migration CVPs, we modified the interactive viewer that we used in Section 8.4. We now have scripts *iviewcvp.scr* and *iviewcvp.sh*. Before we look at the scripts, let's look at the keys (headers) of the output of *migcvp.sh*.

```
        surange  <  migcvp.su
1420 traces:
 tracl=(80,150)  tracr=(1501,3601)  fldr=(1,20)  tracf=(1,2)  cdp=(80,150)
 trid=1 nvs=20 nhs=30 offset=(1200,5000)  sx=(1950,3700)
 gx=(525,2275)  ns=501 dt=4000
```

As expected, the range of *fldr* values is 1-20, there are 1420 traces (20 panels x 71 stack traces (*cdp=80-150*)), and the range of *offset* values (migration velocities) is 1200-5000. Last, key *nvs* has a single value for all traces (20), the number of panels.

9.3.1 Script iviewcvp.scr

We usually start interactive scripts by running a .*scr* script. Here is *iviewcvp.scr*:

```
1  #! /bin/sh
2  # File: iviewcvp.scr
3  #       Run this script to start script iviewcvp.sh
4
5  xterm -geom 80x20+10+545 -e iviewcvp.sh
6
```

Line 1 invokes the shell. Line 5 opens a dialog window 80 characters wide by 20 characters high. Line 5 also places the dialog window 10 pixels from the left side of the viewing area and 545 pixels from the top of the viewing area. Line 5 also starts *iviewcvp.sh*, the processing script, within that window.

9.3.2 Script iviewcvp.sh

We start interactive scripts by running a .*scr* script. Here is *iviewcvp.sh*.

The documentation at the top of the script lists important key information:

- key *fldr* numbers the panels
- key *offset* contains each panel's migration velocity
- key *nvs* contains the total number of panels

Because these keys contain these values, and because the panel number and migration velocity is displayed in each plot, we don't have to remember *migcvp.sh* processing parameters.

Use line 16 to supply the input file name and lines 19-22 to set your plot preferences.

Default settings: The panels are displayed from the first (line 27) to the last. Line 28 is a comment line – the actual last panel value comes from key *nvs*, retrieved by line 52. Panel display increment is "1" (line 29), and the x-axis label is key *cdp* (line 30).

```
 1  #! /bin/sh
 2  # File: iviewcvp.sh
 3  #       View seismic panels from a migration line of
 4  #          Constant Velocity Panels -- key fldr
 5  #       Migration velocity of each panel is in key offset
 6  #       Total number of panels is in key nvs
 7
 8  # Set messages on
 9  ##set -x
10
11  #===============================================
12  # USER AREA -- SUPPLY VALUES
13  #-----------------------------------------------
14
15  # Input seismic data
16  indata=migcvp.su  # SU format
17
18  # Plot choices
19  myperc=98         # perc value for plot
20  plottype=1        # 0 = wiggle plot,  1 = image plot
21  Wplot=300         # Width of plot (pixels)
22  Hplot=500         # Height of plot (pixels)
23
24  #===============================================
25  # Processing variables
26  sortkey=fldr      # sort key [Do Not Change]
27  firsts=1          # first sort (fldr) value:
28  # lasts           # last sort (fldr) value: key nvs
29  increment=1       # sort key (fldr) increment
30  tracekey=cdp      # trace label key
31
32  #-----------------------------------------------
33  # file descriptions
34
35  # tmp1 = binary file of input panel
36  # tmp2 = ASCII file to reduce "zap" screen messages
37
38  #-----------------------------------------------
39
40  echo " "
41  echo "  *** MIGRATION CONSTANT VELOCITY PANEL VIEWER  ***"
42  echo " "
43  echo "       INPUT: $indata"
44
45  #-----------------------------------------------
46  # Remove old temporary files
47  rm -f tmp*
48
49  #-----------------------------------------------
50  # Get first trace - total number of panels in key nvs
51
52  lasts=`sugethw nvs < $indata | sed 1q | sed 's/.*nvs=//'`
53
```

```
 54   #----------------------------------------------------
 55   # BEGIN PANEL LOOP
 56   #----------------------------------------------------
 57
 58   i=$firsts
 59   while [ $i -le $lasts ]
 60   do
 61
 62     suwind < $indata key=$sortkey min=$i max=$i > tmp1
 63
 64   # Retrieve migration velocity from key offset of first trace
 65     migV=`sugethw offset < tmp1 | sed 1q | sed 's/.*offset=//'`
 66     migVel=`expr $migV`  # Ensure migration velocity is numeric
 67
 68     echo " "
 69     echo "Reading panel $i of $indata, Velocity $migVel"
 70     echo "First panel = $firsts    Last panel = $lasts"
 71
 72     if [ $plottype -eq 0 ] ; then
 73       suxwigb < tmp1 xbox=10 ybox=10 wbox=$Wplot hbox=$Hplot \
 74                 title="$sortkey $i    Velocity $migVel" \
 75                 label1=" Time (s)" label2="$tracekey" key=$tracekey \
 76                 perc=$myperc verbose=0 &
 77     else
 78       suximage < tmp1 xbox=10 ybox=10 wbox=$Wplot hbox=$Hplot \
 79                 title="$sortkey $i    Velocity $migVel" \
 80                 label1=" Time (s)" label2="$tracekey" key=$tracekey \
 81                 perc=$myperc verbose=0 &
 82     fi
 83
 84     echo " "
 85     echo "Press Return or Enter to see next panel"
 86     echo "  Or enter \"x\" to exit"
 87     > /dev/tty
 88     read response
 89
 90     case $response in
 91       [xX])
 92             zap xwigb > tmp2
 93             zap ximage > tmp2
 94             i=`expr $lasts + 1`
 95             ;;
 96          *)
 97             zap xwigb > tmp2
 98             zap ximage > tmp2
 99             i=`expr $i + $increment`
100             ;;
101     esac
102
103   done
104
105   #----------------------------------------------------
106   # END PANEL LOOP
107   #----------------------------------------------------
108
109   # Exit
110   echo " "
111   echo "    ==>  Closing iview"
112   pause
113   rm -f tmp*
114   exit
115
```

Let's look at panel loop lines 54-104.

We want to get the offset value, the panel's migration velocity, so we can print it to the screen and put it on the plot. Line 65 is complex; it gets the *offset* value from the first trace of file *tmp1*. The syntax of this line can be used to extract the value of any key from the first trace of *tmp1*. (We saw this syntax in Section 7.6.6 in script *iva.sh*.) Line 62 windows the "current" panel to ensure that line 65 extracts the correct velocity from key *offset*.

Line 66 uses "expr" to ensure that what is extracted from the trace header is purely numeric. In Figure 9.3, the migration velocity printed on the plot includes a strange character. When we include line 66, we do not get the strange character (Figure 9.4).

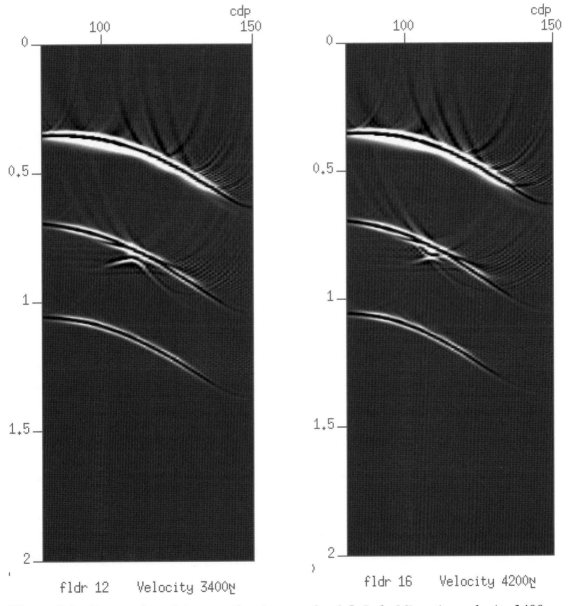

Figure 9.3: *Images from iviewcvp.sh using vscale=1.5. Left: Migration velocity 3400 m/s is too small to collapse the sand channel diffraction. The diffraction is under-migrated. Right: Migration velocity 4200 m/s is too large. Due to over-migration, the*

diffraction is now a "smile." The strange character after "3400" and "4200" is discussed in the text.

Remember, **suximage** reads the first *tracekey* value (here it is *cdp*) and increments by "1" (see Section 7.7). If *tracekey* values do not increment by "1" (for example, due to missing CMPs), the x-axis annotation will be wrong.

In Figure 9.3, notice the many migration "smiles" of the water bottom (1500 m/s rms velocity). These are the result of overmigration (the migration velocity is too high). These figures are designed to find the best value for *vscale* (velocity scale factor) of the deeper sand channel diffraction. (3015 m/s). After using *migcvp.sh* and the viewer, we are satisfied with *vscale=1.9* (Figure 9.4).

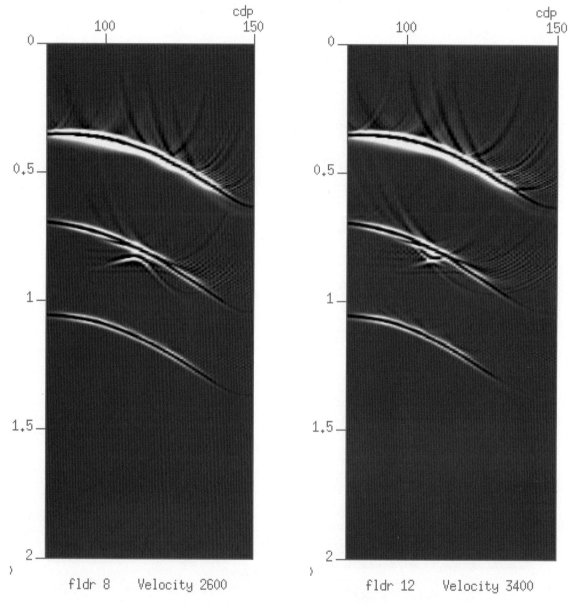

Figure 9.4: *vscale=1.9. Left: Migration velocity for this panel, 2600 m/s, under-migrates the sand channel. Right: Migration velocity for this panel, 3400 m/s, over-migrates the sand channel.*

9.4 Migration Movie

The following script runs a movie of the migration panels. Program **suxmovie** (x-movie plot of a segy data set) and parameter *loop* are the core of the script. We wrote the parameter explanations, lines 12-25, in terms of the migration constant velocity panels. The definitions of these parameters are slightly different in the **suxmovie** and **xmovie** documentation. The following are lines from the **xmovie** selfdoc:

```
Button 1          Zoom with rubberband box
Button 2          reverse the direction of the movie.
Button 3          stop and start the movie.
q or Q key        Quit
s or S key        stop display and switch to Step mode
b or B key        set frame direction to Backward
f or F key        set frame direction to Forward
n or N key        same as 'f'
c or C key        set display mode to Continuous mode
```

If you switch from Movie mode to Step mode by pressing the S key, you can step forward with the F key or backward with the B key. To restart the movie, press the C key.

```
 1   #! /bin/sh
 2   # File: migmovie.sh
 3   #        Run a "movie" of the migration panels
 4   #        Plot "title" shows panel velocity
 5   #        Enter "xmovie" for mouse and keyboard options
 6
 7   set -x
 8
 9   indata=migcvp.su
10   perc=98
11
12   loop=1          # run panels forward continuously
13                   # 2 = run panels back and forth continuously
14                   # 0 = load all panels then stop
15
16   n1=501          # number of time samples
17   d1=0.004        # time sample interval
18   n2=71           # number of traces per panel
19   d2=1            # trace spacing
20
21   width=300       # width of window
22   height=500      # height of window
23
24   fframe=1200     # velocity of first panel for title annotation
25   dframe=200      # panel velocity increment for title annotation
26
27   suxmovie < $indata perc=$perc loop=$loop \
28              n1=$n1 d1=$d1 n2=$n2 d2=$d2 \
29              width=$width height=$height \
30              fframe=$fframe dframe=$dframe \
31              title="Velocity %g" &
32
33   exit
34
```

Parameters *fframe* and *dframe* (lines 24-25) are used to calculate a value for each panel. As the panel (frame) increments, the value of *dframe* is incrementally added to *fframe*. On line 31, "%g" prints the calculated value in the title. If you do not use parameters *fframe* and *dframe*, you can use "%g" to print the frame number in the title.

Observe that the sand channel moves updip as the migration velocity increases.

9.5 Migrate Model 4

The important migration parameters of **sustolt** are *dxcdp*, *tmig*, *vmig*, *smig*, and *vscale*. We calculated *dxcdp* from the stacking chart. Previously, we used **sustolt** as a constant velocity migration program. For that, *tmig* was unimportant, *vmig* had a single value, and *smig* was 1.0 (unused). We used *migcvp.sh* to determine a value for *vscale* based on the known modeled velocity of the sand channel.

Here, we use **sustolt** for 1-D (time-varying) migration in script *migStolt.sh*. Our script has a single time-varying velocity function that is applied across the section, regardless of bed dip. We designed our t-v function to fit the data best near CMP 110, the location of the sand channel diffraction. We selected velocities at different times based on the panels of constant-velocity migrations. We found that a migration velocity at 1500 m/s was appropriate at 0.70 seconds; greater migration velocities provided improved images at later times.

Our migration script, *migStolt.sh*, does not output the migrated data, just a plot (Figure 9.5). By now, you know how to modify the script to produce a data file and make a plot from that data (for example, lines 13-19 of *showshot.sh*, Section 5.6.)

```
 1   #! /bin/sh
 2   # File: migStolt.sh
 3   #        Stolt migration stacked data
 4   #    Input: stack data
 5   #  Output: plot of migrated data
 6   #     Use: migStolt.sh
 7   # Example: migStolt.sh
 8
 9   # Velocities are stacking (Vrms)
10   # Here, we use the false assumption that stacking
11   #     and migrating velocities do not change laterally.
12
13   # smig = stretch factor (0.6 typical if vrms increasing)
14   # vscale = scale factor to apply to velocities
15
16   # Set messages on
17   set -x
18
19   time=0.00,0.70,0.75,0.80,0.85,0.90,2.00
20   vels=1500,1500,2500,3000,2800,3100,3500
21
22   sustolt < stack4.su \
23           cdpmin=1 cdpmax=458 dxcdp=70.71 \
24           tmig=$time vmig=$vels \
25           smig=0.6 vscale=1.9 lstaper=20 lbtaper=100 |
26
27   suximage xbox=10 ybox=10 wbox=800 hbox=400 \
28           title="Migration: Stolt  T = $time  V = $vels" \
29           label1=" Time (s)" label2="CMP" key=cdp \
30           perc=99 verbose=0 &
31
32   # Exit politely from shell
33   exit
34
```

The difference between our two uses of **sustolt** (aside from our use of t-v pairs here) is that we set *smig=0.6* because we are using increasing RMS velocities.

The kink in the second interface at approximately CMP 240 is from the velocity change from 1500 m/s to 2500 m/s at 0.7 seconds. This velocity change works well for the sand channel, but not well at CMP 240. A time-varying and spatially-varying migration would improve the image.

In the pre-migration stack section (Figure 8.3), the sand channel is a strong diffraction and the lower synclines near CMP 180 are unrealistically narrow. After migration (Figure 9.5), the sand channel is greatly collapsed and the lower synclines are realistically gentle.

Careful velocity analysis around CMP 110 (not shown) at the diffractor time of 0.84 seconds produced a stacking velocity of 2670 m/s. It is generally accepted that migration velocities are approximately 10% greater than stacking velocities. 3015 m/s is 13% greater than 2670 m/s.

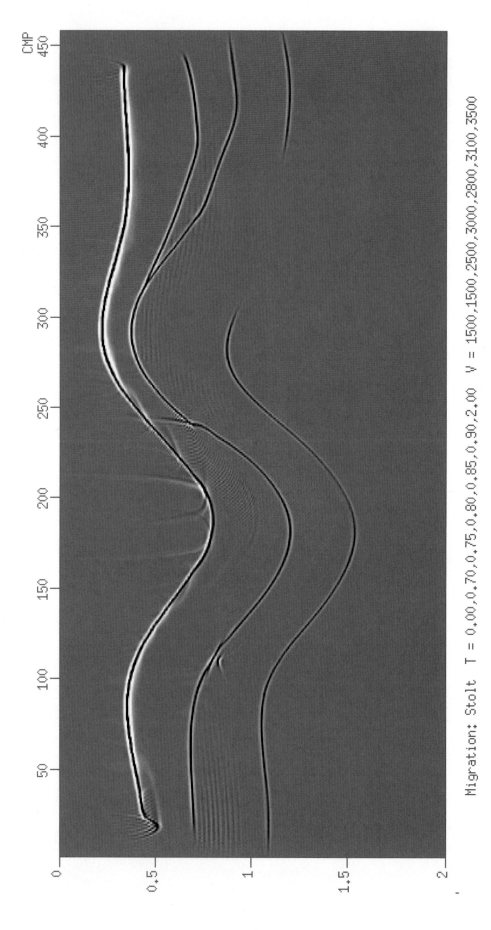

Migration: Stolt   T = 0.00,0.70,0.75,0.80,0.85,0.90,2.00   V = 1500,1500,2500,3000,2800,3100,3500

Figure 9.5: *An **ximage** plot of stack4.su after Stolt migration. Parameter vscale=1.9. The horizontal axis is CMP number. The sand channel diffraction near CMP 110, at approximately 0.8 seconds is greatly reduced (collapsed). Compare to Figure 8.3.*

10. Nankai Data: Examine, Resample, Sort, Gain

10.1 Introduction

We are going to process a real 2-D line of seismic data provided by Prof. Greg Moore of the University of Hawaii. The data were collected near the coast of Japan, over the Nankai trough, where the Philippines plate is subducting beneath Eurasia. The "Nankai" data were collected by the University of Texas, the University of Tulsa, and the University of Tokyo. Based on this data set, a paper was published (Moore, et al., 1990) in which this line is called NT62-8.

The following Nankai seismic files are available:

Table 10.1: Nankai seismic data

Bytes	Name	Description	Traces	Gathers
423827680	Nshots.su	shot-ordered gathers	19057	326
26509560	Nstack.su	stacked data, as published	2869	1

10.1.1 Nankai Shot Gathers

Below is the **surange** output of *Nshots.su*.

```
        surange  <  Nshots.su
19057 traces:
 tracl=(39069,58125)  tracr=(1,19057)  fldr=(1687,2012)  tracf=(28,96)
cdp=(900,1300)
 cdpt=(1,69)  trid=(1,2)  offset=(-2435,-170)  scalel=-10000 scalco=-10000
 counit=1 muts=(0,11000)  ns=5500 dt=2000 day=206
 hour=(21,22)  minute=(0,59)  sec=(1,59)
```

Table 10.2: Some *Nshots.su* key values

Key	Range
fldr	1687 to 2012
cdp	900 to 1300
offset	-2435 to -170
ns	5500
dt	2000

Keys *tracl* and *tracr* number the 19057 traces (although they use different starting numbers).

Key *fldr* tells us there are 326 shot gathers (2012-1687+1).

Since the number of samples per trace (*ns*) is 5500 and the trace sample interval (*dt*) is 2000 microseconds, the trace length is 11 seconds (5500 samples/trace x 0.002 seconds/sample)

Notice that the Nankai shot gather trace headers already contain *cdp* values.

Below is a chart of the number of traces per shot gather (Figure 10.1). (We describe the program, **sukeycount,** which generated the data for the figure in Appendix C.) The first shot gather, 1687, has one trace, then the number of traces per gather increases to a maximum of 69 at gather 1735. This is constant through gather 1965. After gather 1965, the number of traces per gather steadily decreases to 1 at the last gather, 2012.

Nankai Trough

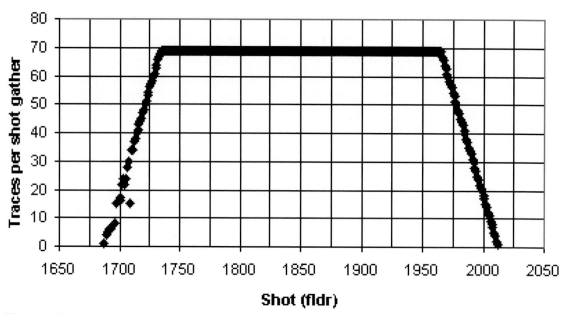

Figure 10.1: *Traces per shot gather in data set Nshots.su.*

10.1.2 Nankai Stack File

Below is the **surange** output of *Nstack.su*. This stack file (Figure 10.2) is the product of processing leading to publication by the University of Hawaii.

```
        surange  <  Nstack.su
2869 traces:
 tracl=(1,2869)  fldr=(0,3383)  tracf=(0,96)  ep=(69,2899)  cdp=(1,2869)
 cdpt=1 trid=1 nhs=(0,72)  offset=(-870,0)  ns=2250
 dt=4000 nofils=(281,3149)  lcf=(69,2899)  hcf=1 lcs=(0,3383)
 hcs=(0,96)  year=(0,207)  hour=(0,23)  minute=(0,59)  sec=(0,59)
 grnors=(0,72)
```

The file of shot gathers, *Nshots.su* contains 401 CMPs (the range of *cdp* values is 900-1300); however, the stack file, *Nstack.su*, contains 2869 traces; in other words, *Nstack.su* is the stack output of 2869 CMPs. This tells us that the shot gather file is a subset of the file that was used by the original processors to make the published stack section.

Table 10.3: Some *Nstack.su* key values

Key	Range
cdp	1 to 2869
ns	2250
dt	4000

File *Nshots.su* has 5500 samples with 2 ms sample interval: 11 seconds of data. In contrast, *Nstack.su* has 2250 traces with 4 ms sample interval: 9 seconds of data. Last, notice that these data were collected in very deep water, approximately 6 seconds two-way time ((6/2) seconds x 1500 meters/second = 4.5 km).

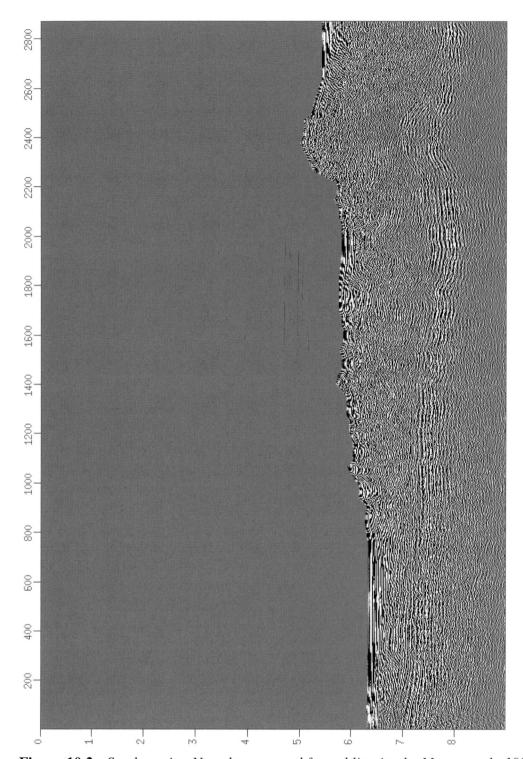

Figure 10.2: *Stack section Nstack.su as used for publication by Moore et al., 1990.*

10.1.3 Our Task

After we examine the shot gathers, we will follow the example of the published data set – we will resample the file from 2 ms to 4 ms sample interval to reduce our prestack processing load by half. Although there does not appear to be any reflection character below eight seconds (see Figure 10.3), we will use all 11 seconds to ensure that we include diffraction limbs for migration.

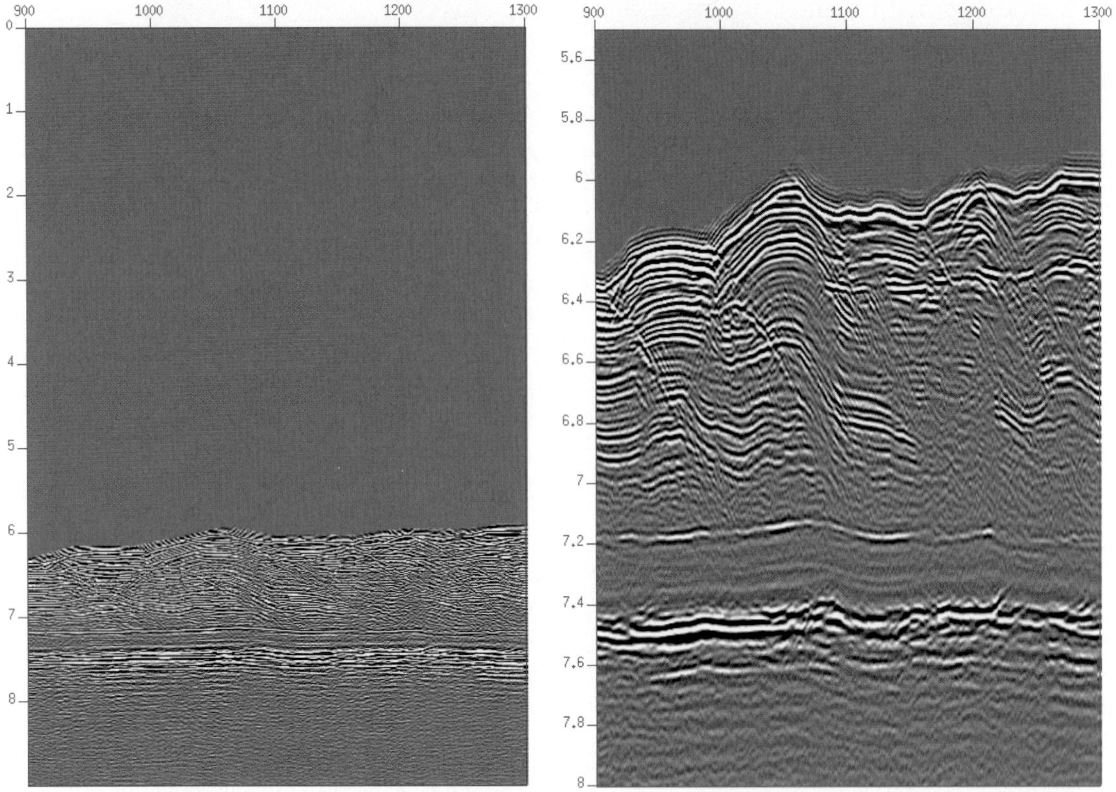

Figure 10.3: *Two views of Nstack.su. Left: CMPs 900-1300. Right: CMPs 900-1300, time 5.5-8 seconds.*

Remember that on CMP gathers, shallow reflections generally exhibit large moveout (large curvature) and deep reflections exhibit little moveout (little curvature). The Nankai data were collected in deep water, so we do not expect to see much moveout on the CMP gathers. Lack of moveout will make velocity analysis difficult, but the quality of the stack is not strongly dependent on the exact velocities used beneath the water bottom.

10.2 Computer Note 12: Symbolic Link

We think these files are too large for everyone to have his or her own copy. Fortunately, there is a simple way for everyone in a group to refer to one copy of the Nankai data: create a symbolic link to the data directory. Suppose the Nankai files are in directory

```
/home/uofhawaii/nankai/Ndata
```

and I (Forel) am in directory

```
/home/forel/suscripts
```

Here, I enter command

```
ln  -s  /home/uofhawaii/nankai/Ndata  .
```

Notice the period at the end. The period means "here" (see Section 6.6). You can read this command as, "Make a symbolic link from ... to ...

```
ln  -s         from              to
```

This command has the effect of placing a virtual subdirectory (*Ndata*) in my directory without actually placing the large data files there. From my *suscripts* directory, I can view a listing of the contents of directory *Ndata* by typing

```
/home/forel/suscripts> ls  -lF  Ndata
total 160
-r--r--r--   1 pennin   geofac   4238276805 May 22 16:46 Nshots.su
-r--r--r--   1 pennin   geofac     26509560 May 22 16:44 Nstack.su
```

Note: To prevent accidentally deleting the data, we recommend you set permissions so the Nankai files are only readable. A brief discussion of permissions is in Section 2.3.5. We also recommend that you store copies of these files in another directory or on a DVD so you can quickly replace deleted files.

10.3   View the Line of Shot Gathers

Before we begin processing the Nankai data, we should examine some of the shot gathers. This is a simple but important quality control (QC) step. Nothing is more important than actually looking at data.

We can use our interactive viewer (Section 8.4). The following lines show how we modified that script for the Nankai data.

```
 8   #==================================================
 9   # USER AREA -- SUPPLY VALUES
10
11   # input seismic data
12   indata=Ndata/Nshots.su  # SU format
13
14   # plot choices
15   myperc=95        # perc value for plot
16   plottype=0       # 0 = wiggle plot,  1 = image plot
17   Wplot=500        # Width of plot (pixels)
18   Hplot=700        # Height of plot (pixels)
19
20   # processing variables
21   sortkey=fldr     # sort key (usually fldr or cdp)
22   firsts=1687      # first sort (fldr or cdp) value
23    lasts=2012      # last sort (fldr or cdp) value
24   increment=20     # sort key increment
25   tracekey=offset # trace label key
26
27   #==================================================
```

One of the important reasons we examine the shot gathers is to determine whether we have bad hydrophones (or for land data, geophones). A bad "phone" is easy to find – usually the entire trace has extremely high amplitudes or zero amplitudes (a "dead" trace). An extremely high amplitude (noisy) trace should usually have its amplitudes replaced by zeros ("killed"); a dead trace usually does not cause processing problems.

If you use the parameters shown above, you will see the images shown in the upper part of Figure 10.4. You might also have to wait a little while for those plots – each time a gather is selected, you have to wait for **suwind** to scroll through the entire 423 Mbytes. An alternative QC procedure is to use a different input file, a time-windowed version of *Nshots.su*. We used the following command:

```
        suwind  <  Nshots.su  tmin=5.8  tmax=7.8  >  NshotsW.su
```

The output file, *NshotsW.su*, occupies only 81 Mbytes. The viewer operates much faster with this smaller data set. (We also changed viewer parameter *Hplot* to 500.)

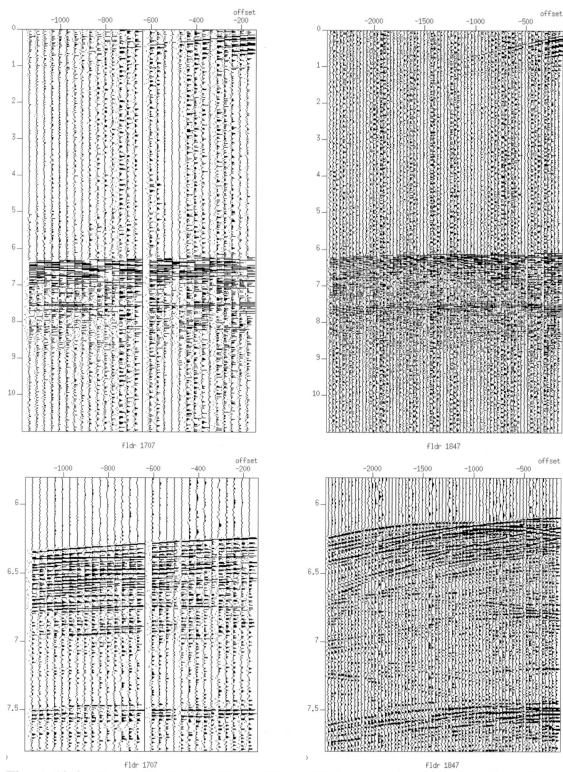

Figure 10.4: *Upper left: Gather 1707, 0-11 sec (30 traces). Upper right: Gather 1847, 0-11 sec (69 traces). Lower left: Gather 1707, 5.8-7.8 sec. Lower right: Gather 1847, 5.8-7.8 sec.*

This data set does not appear to have any noisy traces. (The original processors probably deleted or killed noisy traces. The left plots in Figure 10.4 show a killed trace!)

Note: While the smaller data set is faster to view, you do not see the entire gather. Keep in mind that it is the processor's responsibility to see and comprehend the data. The first time we viewed the Nankai shot gathers, we viewed every 10th gather using all 11 seconds. A problem found early saves much time later!

10.4   f-x Display, Filter, Resample, and Sort Nankai Data

The last time we sorted shot gathers to CMP gathers (Section 7.2), we used script *sort2cmp.sh* to do two tasks:

1. In the file of shot gathers, create header *cdp* and assign values to it.
2. Sort the file of shot gathers to CMP order.

We do not need to do the first task because, as Table 10.2 shows, *Nshots.su* already has *cdp* values. If we only want to do the second task, we would use the command below:

```
susort  <  Nshots.su  >  Ncdps.su  cdp  offset
```

However, we already decided (Section 10.1.2) to change the data sample interval from 2 ms to 4 ms. Below is the documentation for **suresamp**:

```
        sudoc   suresamp
 1   SURESAMP - Resample in time
 2
 3   suresamp <stdin >stdout  [optional parameters]
 4
 5   Required parameters:
 6       none
 7
 8   Optional Parameters:
 9       nt=tr.ns               number of time samples on output
10       dt=tr.dt/10^6          time sampling interval on output
11       tmin=tr.delrt/10^3  first time sample on output
12
13   Example 1: (assume original data had dt=.004 nt=256)
14       sufilter <data f=40,50 amps=1.,0. |
15       suresamp nt=128 dt=.008 | ...
16   Note the typical anti-alias filtering before sub-sampling.
17   Example 2: (assume original data had dt=.004 nt=256)
18       suresamp <data nt=512 dt=.002 | ...
19
20   Credits:
21       CWP: Dave (resamp algorithm), Jack (SU adaptation)
22
23   Trace header fields accessed:  ns, dt, delrt
24   Trace header fields modified:  ns, dt, delrt(only, when set tmin)
25
```

(Notice that instead of simply entering the name of the SU program, we entered "sudoc" and the name of the program. With this, we find that we occasionally unearth a valuable bit of information.)

We placed line numbers on the left for discussion; they are not part of the documentation. Line 16 suggests that we use **sufilter** before we resample the data. The example filter, line 14, begins the high-cut at 40 Hz and by 50 Hz frequencies have zero amplitude.

We accept the recommendation to use an anti-alias filter, a high-cut filter. The alias frequency for 2 ms data is 250 Hz (`1/(2*sample interval)`) and the alias frequency for 4 ms data is 125 Hz. (Two other names for the alias frequency are the

Nyquist frequency and the folding frequency. References: "Frequency Aliasing," Yilmaz, 1987, p. 11 and Yilmaz, 2001, p. 30.) Before we resample our 2 ms data, we need to eliminate frequencies higher than 125 Hz to avoid "folding" them into lower frequencies during resampling.

10.4.1 Script fxdisp

The **suresamp** filter example might be too drastic for our data. That is, we might not want to remove all the frequencies from 50 Hz to the Nyquist frequency of 125 Hz. We can use script *fxdisp.sh* to see a gather and its frequencies.

The *fxdisp.sh* dialog asks for three values: the input file name, a *perc* value for the wiggle plot of the input file, and a key (trace header) to label the x-axis of the wiggle plot. The script creates two displays: a wiggle plot of the input traces (lines 40-44) and a frequency display (lines 33-37). The script transforms each input seismic trace to a frequency trace (line 30); the frequency traces are unrelated to each other. The instruction on line 30

```
suop   op=norm
```

divides each trace by its maximum value. Since this instruction balances each trace, **suximage** (lines 33-37) does not use the *perc* parameter; it is unnecessary.

The script repeats these three questions and two displays until you quit the script. No data are saved after processing.

If you make a mistake while typing, use the Delete key, not the Back Space key.

```
 1  #! /bin/sh
 2  # File: fxdisp.sh
 3  #        Start this script with fxdisp.scr
 4  #
 5  # Transform each input trace to its frequency spectrum
 6  #
 7  #  Inputs (2): gather name, perc, x-axis key
 8  # Outputs (2): f-x display, gather display
 9  #
10  # This is only a display script.  Data are not saved.
11
12  echo " "
13  echo "  *** F-X SPECTRAL DISPLAY ***"
14  echo " "
15
16  #----------------------------------------------
17  # Begin loop
18  #----------------------------------------------
19
20  ok=false
21  while [ $ok = false ]
22  do
23
24    echo "  Enter file name   perc value   key for x-axis"
25    echo "  Example:  dog.su  90  offset"
26    > /dev/tty
27    read indata myperc xkey
28
29  # Create frequency spectrum of each trace and normalize
30    suspecfx < $indata | suop op=norm |
31
32  # Plot frequency traces
33    suximage xbox=10 ybox=10 wbox=400 hbox=600 \
```

```
34              label1=" Freq (Hz)" \
35              windowtitle="f-x spectrum" title=$indata \
36              cmap=hsv7 legend=1 unit=Amplitude \
37              verbose=0 bclip=0.5 wclip=0.0 &
38
39  # Plot input seismic data
40     suxwigb < $indata perc=$myperc key=$xkey \
41              xbox=420 ybox=10 wbox=400 hbox=600 \
42              label1=" Time (s)" label2=$xkey \
43              windowtitle="t-x data" title=$indata \
44              verbose=0 &
45
46  #-------------------------------------------------
47  # Choose loop or exit
48  #-------------------------------------------------
49
50     echo " "
51     echo "  1 = again     2 = exit"
52     > /dev/tty
53     read selection
54       case $selection in
55          1)
56             ok=false
57             ;;
58          2)
59             ok=true
60             zap xwigb > tmp1  # decrease screen messages
61             zap ximage > tmp1  # decrease screen messages
62             ;;
63       esac
64    done
65
66  #-------------------------------------------------
67  # Exit
68  #-------------------------------------------------
69
70  rm -f tmp1
71  exit
72
```

We use script *fxdisp.scr* to run *fxdisp.sh*:

```
1  #! /bin/sh
2  # File: fxdisp.scr
3  #        Run this script to start script fxdisp.sh.
4  #
5  # This is only a display script.  Data are not saved.
6
7  xterm -geom 80x20+10+640 -e fxdisp.sh
8
```

We will not run all the shot gathers through fxdisp; we will window two shot gathers from the 2-D line with the following commands:

```
suwind < Ndata/Nshots.su key=fldr min=1707 max=1707 > Nshot1707.su
suwind < Ndata/Nshots.su key=fldr min=1847 max=1847 > Nshot1847.su
```

Figure 10.5 shows the *fxdisp* plots. Typical seismic data has good high frequency content up to around 50-70 Hz. Frequencies above this are usually noise (unless the survey was designed specifically to record high-frequencies – frequencies up to around 110-150 Hz). Based on these plots, we will design the anti-alias filter to start at 85 Hz and completely zero frequencies at 95 Hz, safely below the 4 ms Nyquist frequency of 125 Hz.

Remember that if you click the middle mouse button in either plot, information of where you click is shown in the upper left corner of the plot. A click in the left plot shows frequency-trace mouse location; a click in the right plot shows time-trace mouse location.

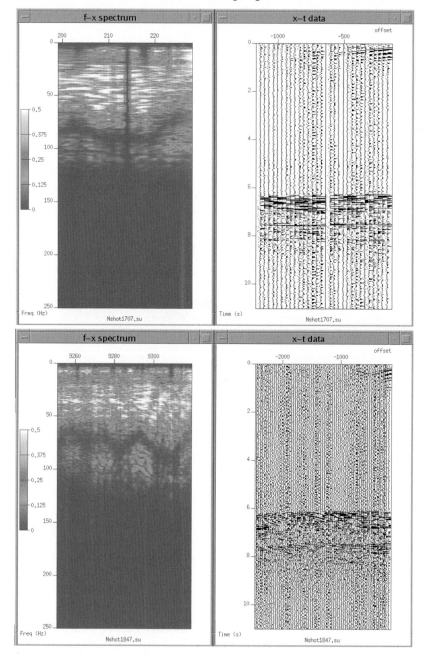

Figure 10.5: *Top: Shot gather 1707, f-x plot and x-t plot. Bottom: Shot gather 1847, f-x plot and x-t plot.*

Also, based on examining the f-x plots of Figure 10.5, we will remove low frequencies that appear to be random signal. We will begin the low-frequency taper at 16 Hz and completely pass frequencies at 21 Hz

The x-axis of the f-x plot is labeled with *tracr* values (the default key); this file does not have valid *f2* key values (see Section 7.7).

10.4.2 Anti-alias Filter, Resample, and Sort

Script, *Nsort2cmp.sh*, filters, resamples, and sorts the shot gathers to CMP gathers.

```
1   #! /bin/sh
2   # File: Nsort2cmp.sh
3
4   # - Filter: cut freq below 21 Hz & above 95 Hz (anti-alias)
5   # - Resample shot gathers from 2 ms to 4 ms sample interval
6   # - Sort to cdp order
7
8     indata=Ndata/Nshots.su
9   outdata=Ndata/Ncdps4.su
10
11  sufilter < $indata f=16,21,85,95 amps=0,1,1,0 |
12  suresamp nt=2750 dt=.004 |
13  susort > $outdata cdp offset
14
15  # Exit politely from shell
16  exit
17
```

Note: The size of *Ncdps4.su* is 214 Mbytes. If you work in a group, we suggest that only one person sort the shot gathers and everyone in the group use a soft link (Section 10.2) to refer to the file of CMP gathers.

Below is the **surange** output of *Ncdps4.su*.

```
        surange  <  Ncdps4.su
19057 traces:
 tracl=(1,19057)  tracr=(1,19057)  fldr=(1687,2012)  tracf=(28,96)  cdp=(900,1300)
 cdpt=(1,69)  trid=(1,2)  offset=(-2435,-170)  scalel=-10000 scalco=-10000
 counit=1 muts=(0,11000)  ns=2750 dt=4000 day=206
 hour=(21,22)  minute=(0,59)  sec=(1,59)
```

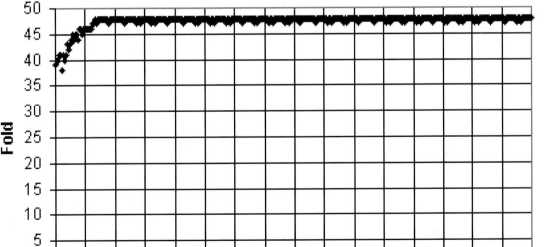

Figure 10.6: *Fold values for data set Ncdps4.su.*

We have 401 gathers. But, the first gathers of the line are not full fold (Figure 10.6). (We describe the program, **sukeycount,** which generated the data for Figure 10.6 in Appendix C.) Upon inspection, we find that the first full-fold CMP gather (48 traces) is CMP 933.

To make reasonably detailed velocity analysis at a regular interval along the line, we choose to analyze every 25th CMP starting with CMP 933; that is, fifteen full-fold CMPs: 933, 958, 983, 1008, … 1208, 1233, 1258, 1283.

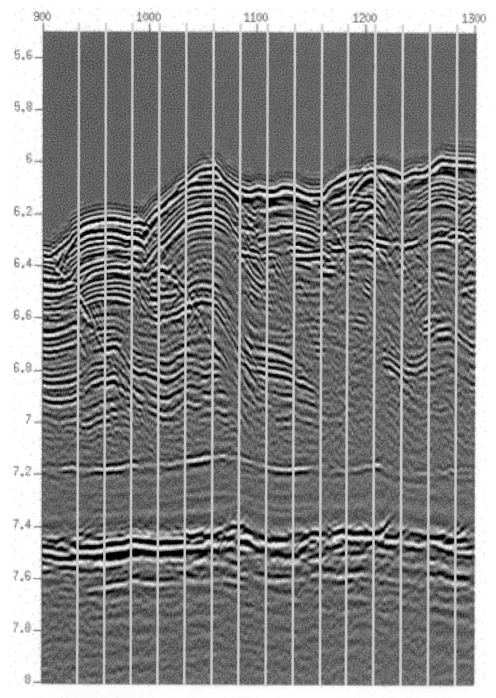

Figure 10.7: *Velocity analysis locations (CMPs) based on Figure 10.3, right.*

10.5   Gain -- Spherical Divergence Correction

To apply spherical divergence correction to the Nankai data, we could simply use "2" in the *tpow* option of **sugain**:

```
sugain  <  Ncdps4.su  tpow=2  >  Ncdps4g.su
```

We might do this because we know that, theoretically, a spherical wavefront weakens as a function of 1/distance squared. However, real data are not always as theory suggests.

To determine which exponential gain to apply to the entire line of CMPs, we will apply gain to two CMPs: 1055 and 1280. We window them from the 2-D line:

```
suwind < Ndata/Ncdps4.su key=cdp min=1055 max=1055 > Ncdp1055.su
suwind < Ndata/Ncdps4.su key=cdp min=1280 max=1280 > Ncdp1280.su
```

Interactive gain script, *igain*, allows us to use most of the options in **sugain**. We can also apply more than one gain to the same file by re-using the gained file.

Script *igain.scr* can be used to run *igain.sh*:

```
1   #! /bin/sh
2   # File: igain.scr
3   #        Run this script to start script igain.sh
4
5   xterm -geom 80x15+10+640 -e igain.sh
6
```

Before running *igain.sh* we supply the name of the seismic file on line 11 and a *perc* value on line 12.

The script creates three displays for the original data and after every time gain is applied (Figures 10.8 and 10.9):

- a wiggle plot of the file,
- a decibel (dB) ximage plot of the user-selected trace(s), and
- an amplitude graph of the user-selected trace(s).

```
1   #! /bin/sh
2   #File: igain.sh
3   #        Run script igain.scr to start this script
4
5   # Set messages on
6   ##set -x
7
8   #================================================
9   # User-supplied values
10
11  indata=Ncdp1280.su  # Input file
12  myperc=95           # perc value
13
14  #================================================
15
16  echo " "
17  echo "  *** INTERACTIVE GAIN TEST ***"
18  echo " "
19
20  # Remove temporary files
21  rm -f tmp*
22
23  #-----------------------------------------------
24  # Get preliminary information: trace(s) to analyze
25  #-----------------------------------------------
26
27  # Ask user for display key
```

```
28  echo " "
29  echo " Supply name of key for trace selection."
30  echo "   For example:  offset  or   tracr"
31  > /dev/tty
32  read mykey
33
34  echo " "
35  echo " Supply minimum and maximum key values for dB amplitude display."
36  echo "   For the best display, use only one to three traces."
37  echo "   for example:  300   300"
38  echo "              or:  250   450"
39  echo "              or: -450  -250"
40  > /dev/tty
41  read mykey1 mykey2
42
43  #-------------------------------------------------
44  # Log preliminary information
45
46  echo "  *** INTERACTIVE GAIN TEST ***" > tmp4
47  echo "Input file = $indata    perc = $myperc" >> tmp4
48  echo "key = $mykey    min value = $mykey1    max value = $mykey2" >> tmp4
49
50  #-------------------------------------------------
51  # Show original gather and spectra first
52  #-------------------------------------------------
53
54  suxwigb < $indata xbox=10 ybox=10 wbox=400 hbox=600 \
55          label1=" Time (s)" label2="$mykey" \
56          title="Original gather" key=$mykey \
57          perc=$myperc verbose=0 &
58
59  suwind < $indata key=$mykey min=$mykey1 max=$mykey2 |
60  suattributes mode=amp |
61  suop op=db > tmp0
62
63  suximage < tmp0 xbox=420 ybox=10 wbox=190 hbox=600 \
64          label1=" Time" label2="Amplitude" title="Amplitude" \
65          grid1=dot grid2=dot legend=1 units=dB \
66          cmap=hsv1 verbose=0 &
67
68  suxgraph < tmp0 -geometry 190x600+620+10 \
69          label1="Time" label2="Amplitude" \
70          title="$mykey $mykey1 $mykey2" grid1=dot grid2=dot \
71          nTic2=2 -bg white verbose=0 &
72
73  #-------------------------------------------------
74  # Amplitude correction
75  #-------------------------------------------------
76
77  new=true  # true = first test
78  ok=false  # false = continue looping
79
80  while [ $ok = false ]
81  do
82
83    rm -f tmp0  # remove earlier copy of file to be gained
84
85    if [ $new = true ] ; then
86      cp $indata tmp0
87      echo " -> Original data" >> tmp4
88    else
89      echo " "
90      echo "Enter A to add another gain correction"
```

```
 91          echo "Enter S to start over"
 92          > /dev/tty
 93          read choice1
 94
 95          case $choice1 in
 96            [sS])
 97                    cp $indata tmp0
 98                    echo " -> Using original data"
 99                    echo " -> Using original data" >> tmp4
100                    ;;
101            [aA])
102                    cp tmp1 tmp0
103                    echo " -> Using modified data"
104                    echo " -> Using modified data" >> tmp4
105                    ;;
106          esac
107
108        fi
109
110        echo " "
111        echo "Select Gain Correction Method:"
112        echo "  Enter A for automatic gain correction"
113        echo "  Enter B to add an overall bias value"
114        echo "  Enter C to clip data"
115        echo "  Enter E to multiply data by exp(t*epow)"
116        echo "  Enter J to use Jon Claerbout values:"
117        echo "            tpow=2  gpow=.5  qclip=.95"
118        echo "  Enter M to balance by dividing by mean"
119        echo "  Enter R to balance data by 1/rms"
120        echo "  Enter S to scale data"
121        echo "  Enter T to multiply data by t^tpow"
122        > /dev/tty
123        read choice2
124
125        case $choice2 in
126          [aA])
127                  echo " Supply window length in seconds:"
128                  > /dev/tty
129                  read wagc
130                  echo " -> AGC: window length = $wagc s"
131                  echo " -> AGC: window length = $wagc s" >> tmp4
132                  sugain < tmp0 agc=1 wagc=$wagc > tmp1
133                  ;;
134          [bB])
135                  echo " Supply overall bias value"
136                  > /dev/tty
137                  read bias
138                  echo " -> Bias with value $bias"
139                  echo " -> Bias with value $bias" >> tmp4
140                  sugain < tmp0 bias=$bias > tmp1
141                  ;;
142          [cC])
143                  echo " Supply clip value between 0.00 and 1.00:"
144                  > /dev/tty
145                  read qclip
146                  echo " -> Clip by $qclip of absolute values"
147                  echo " -> Clip by $qclip of absolute values" >> tmp4
148                  sugain < tmp0 qclip=$qclip > tmp1
149                  ;;
150          [eE])
151                  echo " Supply exponent epow:"
152                  > /dev/tty
153                  read epow
```

```
154                echo " -> Gain function is: A'=A*e^(t*$epow)"
155                echo " -> Gain function is: A'=A*e^(t*$epow)" >> tmp4
156                sugain < tmp0 epow=$epow > tmp1
157                ;;
158        [jJ])
159                echo " -> Jon: tpow=2  gpow=.5  qclip=.95"
160                echo " -> Jon: tpow=2  gpow=.5  qclip=.95" >> tmp4
161                sugain < tmp0 jon=1 > tmp1
162                ;;
163        [mM])
164                echo " -> Balance: divide by mean"
165                echo " -> Balance: divide by mean" >> tmp4
166                sugain < tmp0 mbal=1 > tmp1
167                ;;
168        [rR])
169                echo " -> Balance: divide by rms value"
170                echo " -> Balance: divide by rms value" >> tmp4
171                sugain < tmp0 pbal=1 > tmp1
172                ;;
173        [sS])
174                echo " Supply overall scale factor"
175                > /dev/tty
176                read scale
177                echo " -> Scale with factor $scale"
178                echo " -> Scale with factor $scale" >> tmp4
179                sugain < tmp0 scale=$scale > tmp1
180                ;;
181        [tT])
182                echo " Supply exponent tpow:"
183                > /dev/tty
184                read tpow
185                echo " -> Gain function is: A'=A*t^$tpow"
186                echo " -> Gain function is: A'=A*t^$tpow" >> tmp4
187                sugain < tmp0 tpow=$tpow > tmp1
188                ;;
189    esac
190
191 #-------------------------------------------------
192 # Plot gained data
193 #-------------------------------------------------
194
195    suxwigb < tmp1 xbox=420 ybox=10 wbox=400 hbox=600 \
196            label1=" Time (s)" label2="$mykey" \
197            title="Gain applied" key=$mykey \
198            perc=$myperc verbose=0 &
199
200    suwind < tmp1 key=$mykey min=$mykey1 max=$mykey2 |
201    suattributes mode=amp |
202    suop op=db > tmp2
203
204    suximage < tmp2 xbox=830 ybox=10 wbox=190 hbox=600 \
205            label1=" Time" label2="Amplitude" title="Amplitude" \
206            grid1=dot grid2=dot legend=1 units=dB \
207            cmap=hsv1 verbose=0 &
208
209    suxgraph < tmp2 -geometry 190x600+1030+10 \
210            label1="Time" label2="Amplitude" \
211            title="$mykey $mykey1 $mykey2" grid1=dot grid2=dot \
212            nTic2=2 -bg white verbose=0 &
213
214 #-------------------------------------------------
215 # Choose loop or exit
216 #-------------------------------------------------
```

```
217
218    echo " "
219    echo "Enter 1 for more Amplitude corrections"
220    echo "Enter 2 to output gained seismic data and EXIT"
221    > /dev/tty
222    read choice3
223
224    case $choice3 in
225      1)
226         ok=false
227         ;;
228      2)
229         cp tmp1 igain.su
230         echo " ***  Output data file: igain.su"
231         echo " ***  Output data file: igain.su" >> tmp4
232         cp tmp4 igain.txt
233         echo " ***  Output  log file: igain.txt"
234         pause exit
235         ok=true
236         zap xwigb > tmp3  # decrease screen messages
237         zap ximage > tmp3  # decrease screen messages
238         zap xgraph > tmp3  # decrease screen messages
239         ;;
240    esac
241
242    new=false  # true = first test
243
244  done
245
246  #-----------------------------------------------
247  # Exit
248  #-----------------------------------------------
249
250  # Remove temporary files
251  rm -f tmp*
252
253  # Exit politely from shell
254  exit
255
```

The script applies gain to the entire file, but only the user-selected trace or traces are used to create the second and third displays. In lines 27-41, the user is asked to supply the name of a key (*offset*, *tracr*, etc.) and corresponding key minimum and key maximum values. We selected key *offset* and supplied -1100 -1000 for our offsets. We selected the offsets by plotting the CMP and zooming with the left mouse (Section 1.10).

suxwigb < Ncdp1280.su perc=95 key=offset

We also used **surange** and **sugethw** to examine trace headers:

sugethw < Ncdp1280.su key=tracr,offset,tracf > headers1280.txt

When we looked at our output of **sugethw**, file *headers1280.txt*, we saw that there are two offsets in this range: -1036 and -1003. Two traces are appropriate for this script. We selected offsets away from the direct arrival and that have good reflection character. If, instead, we had selected *tracr* and values 18081 18082, the x-axis values of the wiggle plot would have matched the x-axis values of the dB (second) display (Section 7.7).

Lines 54-71 display the input data. Line 59 windows the selected traces for the second and third display. Line 60 creates the trace attribute of instantaneous amplitude (amplitude envelope traces) and line 61 converts the amplitudes to dB.

We applied (t^pow) gain, (option T), three times, each time independent of the others. (See the log file below.) Table 10.4 shows how the various gain selections enhanced the data's dynamic range.

Table 10.4: dB ranges of Figures 10.8 and 10.9

	No gain	tpow = 1.8	tpow = 2.2	tpow = 2.5
maximum dB	0	0	0	0
minimum dB	-66	-135.6	-157.1	-173.5

Lines 111-189 are where gain is selected and applied. If the user chooses to apply gain again, the user can use the original file (*indata*) or the data that have been gained. Lines 90-106 monitor the choice. The data to be gained are in *tmp0* and after gain the data are in *tmp1*. If the user chooses a fresh gain experiment, line 97 copies the input data, *indata*, to *tmp0*; if the user chooses to gain previously gained data, line 102 copies the gained data, *tmp1*, to *tmp0*.

The script writes the final gained data to disk (line 229 creates file *igain.su*) and a text (ASCII) file of the processing choices (line 232 creates file *igain.txt*). We suggest that you rename the two files to avoid overwriting them by a later run.

Below are the contents of *igain.txt*:

```
 *** INTERACTIVE GAIN TEST ***
Input file = Ncdp1280.su   perc = 95
key = offset    min value = -1100    max value = -1000
 -> Original data
 -> Gain function is: A'=A*t^1.8
 -> Using original data
 -> Gain function is: A'=A*t^2.2
 -> Using original data
 -> Gain function is: A'=A*t^2.5
 ***  Output data file: igain.su
```

To select a *tpow* value to apply to the entire line, we principally rely upon an examination of the third display. We ignore information above six seconds, because that time is when the acoustic energy is in the water layer. Below six seconds, we are looking for a vertical amplitude line. In all three cases (*tpow* = 1.8, 2.2, 2.5), the line is close to vertical. In the third trial, *tpow* = 2.5, the amplitude line is beginning to deviate to the right. After discussion, we decide to use 2.1; it treats the data well and is geophysically reasonable. (CMP 1055 gave similar results.)

Finally, we apply spherical divergence correction with the command that we showed at the beginning of this section, with our new *tpow* value.

```
sugain  <  Ncdps4.su  tpow=2.1  >  Ncdps4g.su
```

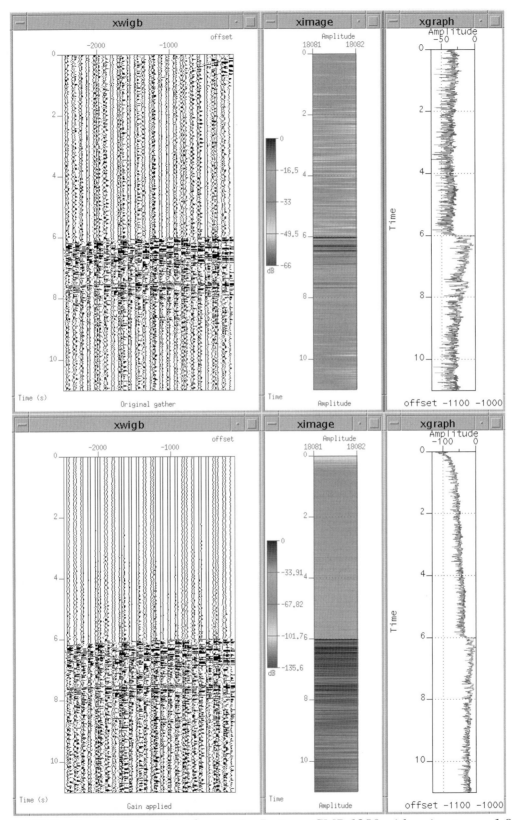

Figure 10.8: *Top: CMP 1280 without gain. Bottom: CMP 1280 with gain, tpow=1.8.*

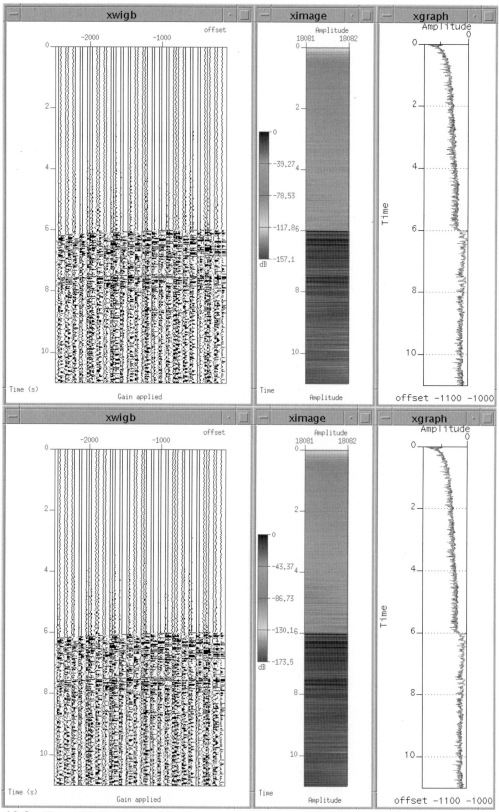

Figure 10.9: *Top: CMP 1280 with gain, tpow=2.2. Bottom: CMP 1280 with gain, tpow=2.5.*

10.6 Single-trace versus Multi-trace Processes

When we consider applying a process, such as spherical divergence correction, should we apply it to the shot gathers or the CMP gathers? It might not matter. If the data order does not matter and we are considering a single-trace process, the result is the same – a trace is processed without regard to its neighbors. Some example SU single-trace processes are listed below.

On the other hand, when we think of applying a multi-trace process, such as NMO, the order of the data is important. Changing the order of the data (shot gathers versus CMP gathers versus a stack file) dramatically changes the output. Some examples of SU multi-trace processes are listed below.

Table 10.5: Compare processes

Single-trace	Multi-trace
sugain – gain	sunmo – NMO
sufilter – 1-D filter	suvelan – semblance
sukill – kill trace	sudipfilt – 2-D filter
supef – deconvolution	sustolt – Stolt migration

A further question is, why are we thinking of applying a process? Spherical divergence correction is a scientifically valid way to balance a trace in time. We need this to re-balance data that are naturally attenuated in time. Therefore, we apply this gain early in the processing stream.

Normal moveout correction, NMO, is meaningful to CMP gathers, but not shot gathers. On the other hand, we might apply deconvolution, a single-trace process, twice. The first time might be a gap deconvolution to remove multiples; the second application would be a spiking deconvolution to sharpen the reflections (compress the wavelets).

10.7 Summary

This chapter saw us working with large files and real data. When we processed model data, we went easily from shot gathers to sort to CMP gathers to velocity analysis. Real data require inspection, gain, frequency analysis, and much more that we will see in later chapters.

11. Nankai: Velocity Analysis, NMO, Stack, Migration

11.1 Velocity Analysis

Below is the user area of *iva.sh* changed for the Nankai data. As stated in Section 10.4.2, we will do velocity analysis on every 25th CMP starting with CMP 933 (lines 13-16). We change the *perc* value to 95 (line 29), an accommodation for real data. Semblance values go from 1000 m/s (line 38) to 7000 m/s (nvs*dvs + fvs). CVS panels range from 1000 m/s to 6000 m/s (lines 41-44).

```
 8   #================================================
 9   # USER AREA -- SUPPLY VALUES
10   #------------------------------------------------
11   # CMPs for analysis
12
13    cmp1=933    cmp2=958    cmp3=983    cmp4=1008
14    cmp5=1033   cmp6=1058   cmp7=1083   cmp8=1108
15    cmp9=1133  cmp10=1158  cmp11=1183  cmp12=1208
16   cmp13=1233  cmp14=1258  cmp15=1283
17
18   numCMPs=15
19
20   #------------------------------------------------
21   # File names
22
23   indata=Ndata/Ncdps4g.su  # SU format
24   outpicks=Nvpick.txt  # ASCII file
25
26   #------------------------------------------------
27   # display choices
28
29   myperc=95        # perc value for plot
30   plottype=1       # 0 = wiggle plot,  1 = image plot
31
32   #------------------------------------------------
33   # Processing variables
34
35   # Semblance variables
36   nvs=240  # number of velocities
37   dvs=25   # velocity intervals
38   fvs=1000 # first velocity
39
40   # CVS variables
41   fc=1000 # first CVS velocity
42   lc=6000 # last CVS velocity
43   nc=10   # number of CVS velocities (panels)
44   XX=11   # ODD number of CMPs to stack into central CVS
45
46   #================================================
```

By looking at the Nankai stack, Figure 10.3 left, we know the water layer is approximately 6 seconds (two-way time) and we think there is not much reflection character below 7.8 seconds. The left side of Figure 11.1 is the velan of CMP 933. It appears that the time for picking events is restricted to between 6 seconds and 8 seconds.

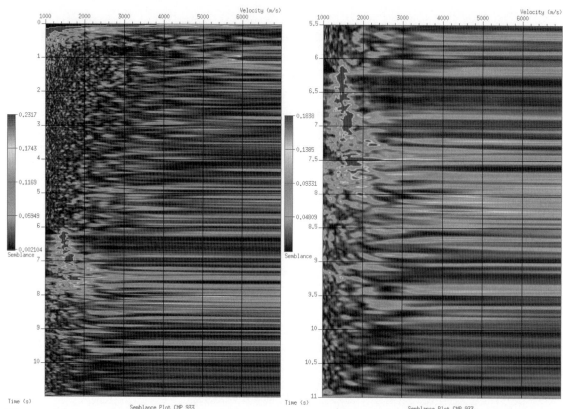

Figure 11.1: *Left: Velan of CMP 933, 0-11 seconds. Right: Velan of CMP 933, 5.5-11 seconds.*

Instead of making velocity picks in a narrow space on the screen, we window the file of CMPs:

```
suwind < Ncdps4g.su tmin=5.5 tmax=11 > Ncdps4g5511.su
```

and use this as the input to iva.sh

```
23   indata=Ndata/Ncdps4g5511.su  # SU format
```

We chose the full time (*tmax=11*) because on CMP 933 we see events around 9.5 seconds and we don't want to miss these and other potentially later events.

The result of using this new input file is shown on the right side of Figure 11.1. This makes better use of the screen.

Because we are not picking from zero time, we will have to modify the output pick file, adding zero time to the beginning of all the *tnmo* series and adding the water velocity to the beginning of all the *vnmo* series.

Below is the output file, *Nvpick.txt*. For the sake of presentation, we made line 1 into two lines. Line 1 as presented here has no spaces at the end of the first part and no spaces at the beginning of the second part, making this a usable file.

```
1   cdp=933,958,983,1008,1033,1058,1083,\
    1108,1133,1158,1183,1208,1233,1258,1283 \
2   #=1,2,3,4,5,6,7,8,9,10,11,12,13,14,15 \
3   tnmo=5.52263,6.44307,6.97119,7.65775,8.50274,10.9849 \
4   vnmo=1522.81,1460.57,1622.4,1834.01,2331.93,2892.08 \
5   tnmo=5.53018,6.49588,7.55967,9.19685,10.9774 \
6   vnmo=1497.92,1497.92,1634.84,2966.77,3016.56 \
7   tnmo=5.52263,6.35254,7.40878,9.27229,10.9849 \
```

```
 8  vnmo=1497.92,1460.57,1659.74,2307.03,2991.67 \
 9  tnmo=5.53772,7.48422,8.93279,10.9925 \
10  vnmo=1497.92,1634.84,2282.14,3564.27 \
11  tnmo=5.55281,6.99383,7.5144,8.48011,10.9925 \
12  vnmo=1522.81,1659.74,1734.43,2319.48,2991.67 \
13  tnmo=5.53018,6.69959,7.46159,9.45336,10.9774 \
14  vnmo=1485.47,1572.6,1871.35,2282.14,2929.43 \
15  tnmo=5.53018,6.79012,7.58985,8.88752,10.9774 \
16  vnmo=1497.92,1622.4,1771.77,2406.61,2979.22 \
17  tnmo=5.53772,6.29973,7.3786,8.7668,10.9698 \
18  vnmo=1497.92,1497.92,1759.32,2842.29,3589.17 \
19  tnmo=5.53772,7.30315,9.16667,10.9925 \
20  vnmo=1510.36,1672.19,2630.68,3053.91 \
21  tnmo=5.52263,7.39369,9.60425,10.9849 \
22  vnmo=1497.92,1721.98,2443.96,3004.11 \
23  tnmo=5.53018,6.93347,7.55213,10.9849 \
24  vnmo=1535.26,1634.84,1734.43,2867.19 \
25  tnmo=5.52263,7.43896,8.2915,10.9849 \
26  vnmo=1510.36,1647.29,2282.14,2979.22 \
27  tnmo=5.53018,7.49931,8.13306,10.9698 \
28  vnmo=1510.36,1659.74,1921.15,2555.99 \
29  tnmo=5.54527,7.35597,10.9849 \
30  vnmo=1510.36,1697.08,3004.11 \
31  tnmo=5.53772,7.29561,8.19342,9.08368,10.9925 \
32  vnmo=1522.81,1746.88,2456.41,2879.64,4248.91 \
```

Our picking philosophy was simple: slowly increasing velocity with depth. We did not consider any picks that deviated strongly from this. We think this is adequate for a first look at stack and migration. (In the beginning, simple is good.) After we examine the stack data and the migrated data, we can reconsider our picks.

The file *Nvpick.txt* must be modified before it can be used for NMO:

- Line 2 must be removed because a comment line cannot be in the midst of a command. The continuation mark "\" continues a command onto the next line. (However, a comment line can be between piped (|) commands.)

- We picked from a seismic file that did not go to zero time, so we must add the zero time and the velocity at zero time to the *tnmo* and *vnmo* lines. The result is in the script in the next section.

Note: If we had windowed the input file to exclude the last time; for example, if we had started the window at 5.5 seconds and ended the window at 10 seconds

```
suwind  <  Ncdps4g.su  tmin=5.5  tmax=10  >  Ncdps4g5510.su
```

we would have to edit *Nvpick.txt* at the start of each series and at the end of each series. At the end of each *tnmo* series, we would add 11.0 and at the end of each vnmo series we would repeat the last picked velocity.

Because of the rough sea-floor topography and the spatial-temporal interpolation of the velocity profiles, it may be wise to pick velocities at much more closely spaced CMP locations. We leave this exercise to the reader who should be able to improve on our final image by performing this additional task.

11.2 NMO and Stack

File *Nnmo.sh* is applies NMO and stack to the 2-D line of gained CMPs. The size of the stack file is only 4.5 Mbytes. Notice that the *tnmo* lines have 0.0 as the first value and the *vnmo* lines have 1500. as the first value.

For the sake of presentation, we made line 18 into two lines. Line 18 as presented here has no spaces at the end of the first part and no spaces at the beginning of the second part, making this a usable file.

```
 1   #! /bin/sh
 2   # File: Nnmo.sh
 3   #       Apply NMO (flatten) 2-D line of CMPs
 4   # Input (1): 2-D line of CMPs
 5   # Output (1): NMO-corrected 2-D line of CMPs
 6   # Use: Nnmo.sh
 7   #
 8   # NMO correction is interpolated between named CMPs.
 9
10   # Set debugging on
11   set -x
12
13   # Name data sets
14    indata=Ndata/Ncdps4g.su
15   outdata=Nstack4.su
16
17   sunmo < $indata \
18   cdp=933,958,983,1008,1033,1058,1083,\
     1108,1133,1158,1183,1208,1233,1258,1283 \
19   tnmo=0.0,5.52263,6.44307,6.97119,7.65775,8.50274,10.9849 \
20   vnmo=1500.,1522.81,1460.57,1622.4,1834.01,2331.93,2892.08 \
21   tnmo=0.0,5.53018,6.49588,7.55967,9.19685,10.9774 \
22   vnmo=1500.,1497.92,1497.92,1634.84,2966.77,3016.56 \
23   tnmo=0.0,5.52263,6.35254,7.40878,9.27229,10.9849 \
24   vnmo=1500.,1497.92,1460.57,1659.74,2307.03,2991.67 \
25   tnmo=0.0,5.53772,7.48422,8.93279,10.9925 \
26   vnmo=1500.,1497.92,1634.84,2282.14,3564.27 \
27   tnmo=0.0,5.55281,6.99383,7.5144,8.48011,10.9925 \
28   vnmo=1500.,1522.81,1659.74,1734.43,2319.48,2991.67 \
29   tnmo=0.0,5.53018,6.69959,7.46159,9.45336,10.9774 \
30   vnmo=1500.,1485.47,1572.6,1871.35,2282.14,2929.43 \
31   tnmo=0.0,5.53018,6.79012,7.58985,8.88752,10.9774 \
32   vnmo=1500.,1497.92,1622.4,1771.77,2406.61,2979.22 \
33   tnmo=0.0,5.53772,6.29973,7.3786,8.7668,10.9698 \
34   vnmo=1500.,1497.92,1497.92,1759.32,2842.29,3589.17 \
35   tnmo=0.0,5.53772,7.30315,9.16667,10.9925 \
36   vnmo=1500.,1510.36,1672.19,2630.68,3053.91 \
37   tnmo=0.0,5.52263,7.39369,9.60425,10.9849 \
38   vnmo=1500.,1497.92,1721.98,2443.96,3004.11 \
39   tnmo=0.0,5.53018,6.93347,7.55213,10.9849 \
40   vnmo=1500.,1535.26,1634.84,1734.43,2867.19 \
41   tnmo=0.0,5.52263,7.43896,8.2915,10.9849 \
42   vnmo=1500.,1510.36,1647.29,2282.14,2979.22 \
43   tnmo=0.0,5.53018,7.49931,8.13306,10.9698 \
44   vnmo=1500.,1510.36,1659.74,1921.15,2555.99 \
45   tnmo=0.0,5.54527,7.35597,10.9849 \
46   vnmo=1500.,1510.36,1697.08,3004.11 \
47   tnmo=0.0,5.53772,7.29561,8.19342,9.08368,10.9925 \
48   vnmo=1500.,1522.81,1746.88,2456.41,2879.64,4248.91 |
49   sustack > $outdata
50
51   # Exit politely from shell
52   exit
53
```

We plot the stack file with **suximage**.

```
suximage  <  Nstack4.su  key=cdp  perc=95  &
```

We see that our stack file (Figure 11.2) does not resemble the original stack file (Figure 10.3, right). We suspect the original file *Nstack.su* is actually a stacked and migrated file. (In fact, this is the case, as is clear from the publication by Moore et al., 1990.)

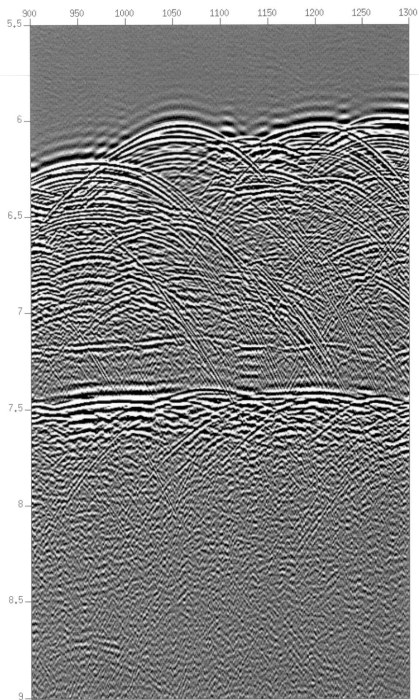

Figure 11.2: *Stack data set Nstack4.su.*

11.3   Stolt Migration with sustolt

11.3.1 Script Nmigcvp.sh

Below is the user area of *Nmigcvp.sh* for the Nankai data. The CDP bin distance (reported in Moore, et al., 1990) is 16.667 m (parameter *dxcdp*, line 22). Because we do not know rms velocities for these data, we set *vscale=1.0* (line25). We process for image, not for velocities. Our first velocity is 1400 m/s and our last is 2000 m/s, a narrow range.

```
11   #================================================
12   # USER AREA -- SUPPLY VALUES
13   #------------------------------------------------
14
15   # Seismic files
16   indata=Nstack4.su    # SU format
17   outdata=Ndata/Nmigcvp.su    # migration Constant Velocity Panels
18
19   # Migration variables
20   cdpmin=900       # Start CDP value
21   cdpmax=1300      # End CDP value
22   dxcdp=16.667     # distance between adjacent CDP bins (m)
23   smig=1.0         # stretch factor (0.6 typical if vrms increasing)
24                    # [the "W" factor] (Default=1.0)
25   vscale=1.0       # scale factor to apply to velocities (Default=1.0)
26   lstaper=20       # length of side tapers (traces) (Default=0)
27   lbtaper=100      # length of bottom taper (samples) (Default=0)
28
29   # Velocity panel variables
30   firstv=1400      # first velocity value
31    lastv=2000      # last velocity value
32   increment=200    # velocity increment
33
34   numVtest=100     # use to limit number of velocity panels
35                    # otherwise, use very large value (100)
36
37   #================================================
```

11.3.2 Script Niviewcvp

Below is the user area of *Niviewcvp.sh* for the Nankai data. We set *Wplot* and *Hplot* to accommodate a larger image.

```
11   #================================================
12   # USER AREA -- SUPPLY VALUES
13   #------------------------------------------------
14
15   # Input seismic data
16   indata=Nmigcvp.su  # SU format
17
18   # Plot choices
19   myperc=95        # perc value for plot
20   plottype=1       # 0 = wiggle plot,  1 = image plot
21   Wplot=550        # Width of plot (pixels)
22   Hplot=700        # Height of plot (pixels)
23
24   #================================================
```

We also changed line 62. For Model 4, the line is:

```
62     suwind < $indata key=$sortkey min=$i max=$i > tmp1
```

To view the Nankai data, we also window time:

```
62     suwind < $indata key=$sortkey min=$i max=$i tmin=5.5 tmax=8 > tmp1
```

We do not want to waste viewing space on the deep-water layer, and we see no significant reflections below 7.8 seconds.

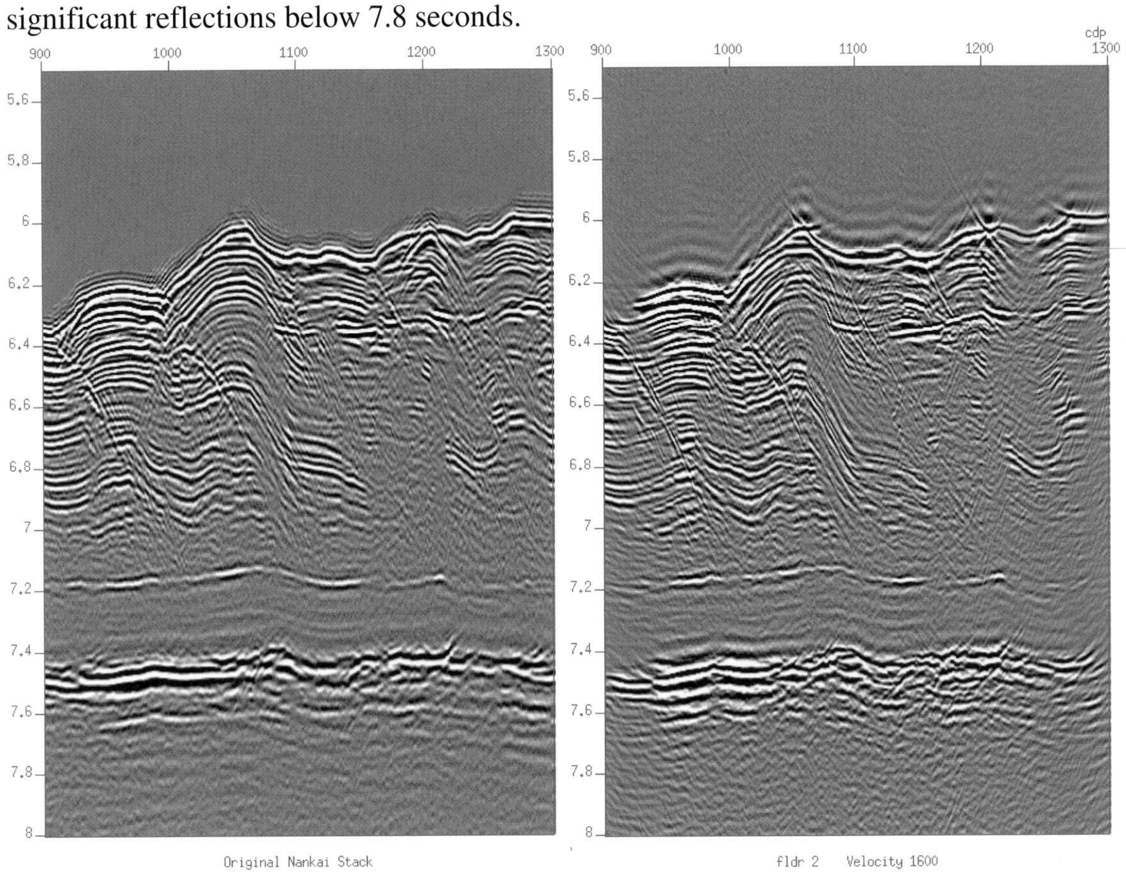

Figure 11.3: *Left: Original migrated data file. Right: Our constant-velocity Stolt migration at 1600 m/s.*

We find that our best match to the original data, and our best image, is the 1600 m/s migration (Figure 11.3). Notice that the migration "smiles" above the water bottom have been muted on the original data set, cosmetically improving the data set's appearance.

11.3.3 Script Nmigmovie.sh

Below is script *Nmigmovie.sh* for the Nankai data. We set *loop=2* (line 12) to run the movie back and forth. Below, *fframe=1300* and *dframe=100* because we re-ran the migration (*Nmigcvp.sh*) for velocities 1300-2000 m/s at 100 m/s increment.

```
 1   #! /bin/sh
 2   # File: Nmigmovie.sh
 3   #        Run a "movie" of the migration panels
 4   #        Plot "title" shows panel velocity
 5   #        Enter "xmovie" for mouse and keyboard options
 6
 7   set -x
 8
 9   indata=Ndata/Nmigcvp.su
10   perc=98
11
12   loop=2          # 1 = run panels forward continuously
13                   # 2 = run panels back and forth continuously
14                   # 0 = load all panels then stop
15
```

```
16   n1=2750      # number of time samples
17   d1=0.004     # time sample interval
18   n2=401       # number of traces per panel
19   d2=1         # trace spacing
20
21   width=550    # width of window
22   height=700   # height of window
23
24   fframe=1300  # velocity of first panel for title annotation
25   dframe=100   # panel velocity increment for title annotation
26
27   suxmovie < $indata perc=$perc loop=$loop \
28               n1=$n1 d1=$d1 n2=$n2 d2=$d2 \
29               width=$width height=$height \
30               fframe=$fframe dframe=$dframe \
31               title="Velocity %g" &
32
33   exit
34
```

You can zoom the xmovie window the same way you zoom the other plot windows (Section 1.10).

This movie also shows that, using a single migration velocity, 1600 m/s is a good match for the original data. Note that this is just a small amount faster than the speed of sound in the water layer, suggesting that *vscale=1.0* is appropriate, and that the majority of migration is due to propagation in the water layer.

We then used a script similar to migStolt.sh (see Section 9.5) to migrate and display the data set. If we were satisfied with a single migration velocity, we might use a line command (for an example, see Section 11.4).

11.3.4 Summary

The Nankai data shows many diffractions. As you look at the various migrations, you can see that reflectors do not change position; this is generally flat geology. When reflectors are under-migrated, the diffractions are not collapsed – they are "frowns." When reflectors are over-migrated, the diffractions change to "smiles." The original data set appears to be slightly over-migrated.

After we migrated Model 4 (in Chapter 9), we used CMP 110 to compare the stacking velocity to the migration (rms) velocity. That was a valid comparison because we had calibrated the velocity scale factor, *vscale* using our model velocity. However, we cannot expect to derive geologic velocities by migrating the Nankai data because we did not do a similar calibration.

The Nankai data set is difficult to process because the geology is under 4.5 km of water. Because of this, the moveout that we expect to see and use for velocity analysis is subtle. Proper velocity analysis of this data set demands patience and some prior knowledge of the local velocities, particularly the water velocity. We encourage you to improve upon our image.

11.4   *Phase Shift Migration with sugazmig*

Another migration program is **sugazmig**, an SU version of Jeno Gazdag's phase-shift migration for zero-offset data. The following migration uses a single time-velocity function. (Program **sumigpspi** is an SU version of J. Gazdag's phase-shift plus interpolation migration for zero-offset data. It can handle lateral velocity variations.)

Program **sugazmig** is easy to use – it gets most of the information that it needs from the trace keys. Aside from the t-v pairs, the only parameter value that we supply below is for *dx*, **sugazmig**'s name for the CDP bin distance. The velocities are an approximation based on the previous Stolt migrations.

Note: For **sugazmig**, the velocities are *INTERVAL VELOCITIES*.

```
 1  #! /bin/sh
 2  # File: Ngazmig.sh
 3  #       Phase shift migration of Nankai stacked data.
 4  #       CDP spacing (dx) is 16.667 meters.
 5
 6  # Set messages on
 7  set -x
 8
 9  sugazmig < Nstack4.su dx=16.6667                    \
10            tmig=0,5,6,7,8,9                          \
11            vmig=1500,1500,1950,2000,2050,2100 \
12          > Ngazmig.su
13
14  # Exit politely from shell
15  exit
16
```

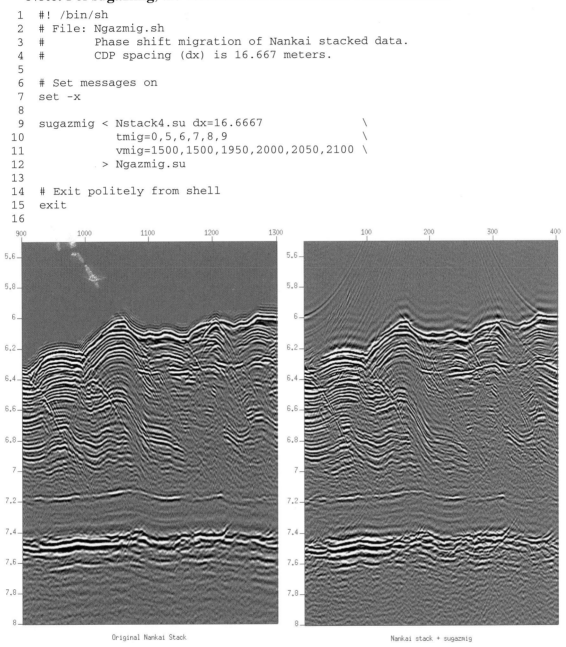

Original Nankai Stack   Nankai stack + sugazmig

Figure 11.4: *Left: Original migrated data file. Right: Our migration using **sugazmig**.*

This migration took half an hour to run on a Sun UltraSPARC III with four processors. Figure 11.4 shows that our migration is a fair match to the original migration.

Although **sugazmig** does some internal zero-padding, you can see minor migration artifacts on the sides of the section at about 6 seconds. If you look at the bottom of the time section (11 seconds), you can also see migration artifacts. However, the bottom artifacts are unimportant since there are no strong reflections below 10 seconds. Program **sugazmig** does not have taper parameters; in contrast to **sustolt** parameters *lstaper* and *lbtaper*.

We used the following command to display the migrated file:

```
suwind < Ngazmig.su tmin=5.5 tmax=8 | suximage perc=95 title="Nankai stack +
sugazmig" &
```

11.5   Some Geologic Features

Figure 11.5 (below) shows a time-windowed portion of the original stacked and migrated data set

Oceanic crust of the Pacific Plate lies under a veneer of sediments at the left side of the image, and is subducted at the trench, which is filled with recent sediments (CDPs 200-600, approximately). To the right of these flat-lying sediments, the accretionary prism is evident, the distorted sediment layer thickens to the right, and some extreme topography results.

11.5.1 Bottom-simulating Reflector

Notice the reflector that seems to mimic the ocean-bottom topography, but is about 300 ms later. It cuts across the folded sediments, and therefore cannot be representative of a depositional boundary. This bottom-simulating reflector (BSR) is a common feature in some basins. It represents a reflection from a change in fluid phases; in most cases it corresponds to the base of a layer of methane hydrate, a gas trapped in a frozen state within an ice "foam." Beneath the hydrate, the gas exists in a free state. Temperature (and pressure) conditions dictate where this transition occurs. Because the sea floor is nearly a constant-temperature boundary, the phase transition occurs at a more-or-less constant depth (locally) beneath the sea floor.

11.5.2 Subducted Pacific Plate

Notice that the Pacific Plate oceanic crust appears to flatten at about 7.2 seconds, rather than continue to increase "depth" to the right of the trench. This is an artifact of velocity pull-up. The deformed sediment layer with seismic velocities of about 2000 m/s replaces water with velocities of 1500 m/s. Thus, an increasing time between the water bottom and the subducted crust implies increasing depth to the reflection, even though its reflection time happens to be nearly constant.

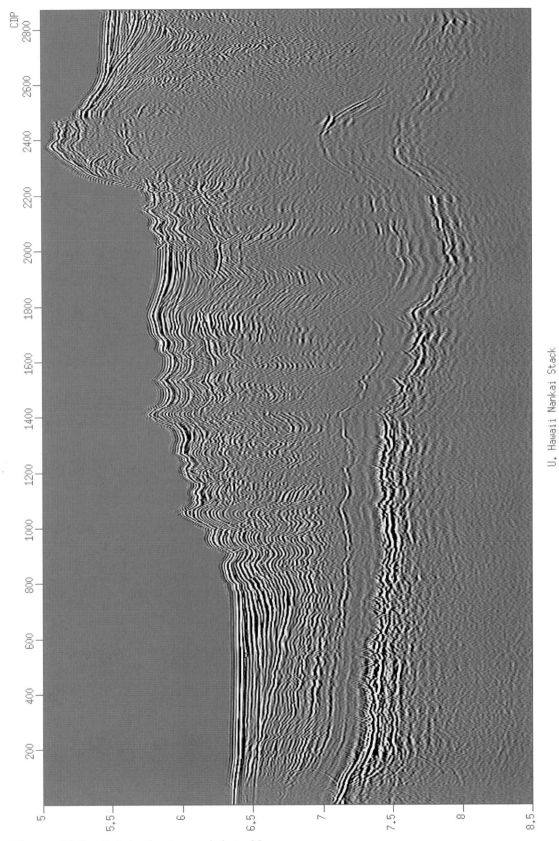

U. Hawaii Nankai Stack

Figure 11.5: *Original migrated data file.*

12. Taiwan Data: Examine, Zero Traces, Re-examine

12.1 Introduction

Our third data set is another 2-D line of seismic data provided by Prof. Greg Moore of the University of Hawaii. This "Taiwan" data set was collected near the coast of Taiwan in 1995 (Berndt & Moore, 1999) by the University of Hawaii, San Jose State University, and National Taiwan University. The size of our shot gather file, *Tshot.su*, is 411 Mbytes. (We have only one Taiwan data set, the original shot gathers.) This data set is more difficult to process than the Nankai data set.

Below is the **surange** output of *Tshot.su*.

```
        surange  <  Tshot.su
25344 traces:
 tracl=(114769,140112)  tracr=(1,25344)  fldr=(800,975)  tracf=(1,144)
ep=(740,915)
 cdp=(4027,4816)  cdpt=(1,144)  trid=1 nhs=1 offset=(-3663,-88)
 sdepth=80000 swdep=(2120000,3540000)  scalel=-10000 scalco=1 sx=(-4773100,-
3965250)
 gx=(-4764600,-3600343)  gy=(766,94805)  counit=3 tstat=12 ns=3999
 dt=4000 gain=9 afilf=160 afils=72 lcf=3
 hcf=160 lcs=6 hcs=72 year=95 day=260
 hour=(12,13)  minute=(0,59)  sec=(0,59)
```

Table 12.1: Some *Tshot.su* key values

Key	Range
fldr	800 to 975
cdp	4027 to 4816
offset	-3663 to -88
ns	3999
dt	4000

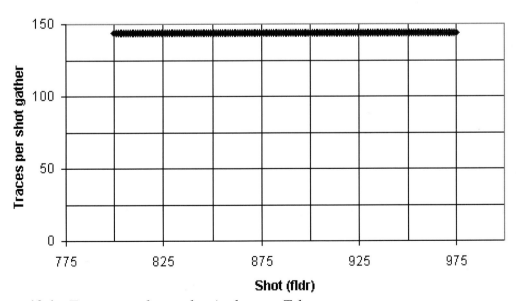

Figure 12.1: *Traces per shot gather in data set Tshot.su.*

Keys *tracl* and *tracr* number the 25344 traces (although, they use different starting numbers).

Key *fldr* tells us there are 176 shot gathers (975-800+1).

Since the number of samples per trace (*ns*) is 3999 and the trace sample interval (*dt*) is 4000 microseconds, the trace length is 16 seconds (15.996 seconds)

Notice that the Taiwan shot gather trace headers already contain *cdp* values.

This data set has many acquisition key values such as source x-coordinate (*sx*), receiver x-coordinate (*gx*), receiver y-coordinate (*gy*), and source depth (*sdepth*).

Figure 12.1 is a chart of the number of traces per shot gather. (We describe the program, **sukeycount**, which generated the data for the figure in Appendix C.) The chart shows that every shot gather has 144 traces.

We do not want to process 16 seconds of data. Let's use the first 5 seconds:

```
suwind  <  Tshot.su  tmax=5  >  Tshot5.su
```

The size of our windowed shot gather file is 133 Mbytes.

12.2 Reminder: Symbolic Link

As we discussed in Section 10.2 for the Nankai data, we think this Taiwan shot file is too large for everyone to have his or her own copy. We suggest everyone in a group create a symbolic link to the data directory. Suppose the Taiwan file is in directory

```
/home/uofhawaii/taiwan/Tdata
```

and I (Forel) am in directory

```
/home/forel/suscripts
```

Here, I enter command

```
ln  -s  /home/uofhawaii/taiwan/Tdata  .
```

Remember, the period at the end represents "here." This command makes a symbolic link from ... to ...

```
ln  -s            from            to
```

This command has the effect of placing a virtual subdirectory (*Tdata*) in my directory.

To prevent accidentally deleting the data, we recommend you set permissions so the Taiwan file is only readable. We also recommend that you store a copy of this file in another directory or on a DVD so you can quickly replace the deleted file.

12.3 View the Line cf Shot Gathers

Before we begin processing the Taiwan data, we should examine some of the shot gathers. This is a simple but important quality control (QC) step. Nothing is more important than actually looking at data.

We can use our interactive viewer (Section 8.4). The following lines show how we modified that script for the Taiwan data.

```
 8   #================================================
 9   # USER AREA -- SUPPLY VALUES
10
11   # input seismic data
12   indata=Tdata/Tshot5.su  # SU format
13
```

```
14   # plot choices
15   myperc=95        # perc value for plot
16   plottype=0       # 0 = wiggle plot,  1 = image plot
17   Wplot=900        # Width of plot (pixels)
18   Hplot=600        # Height of plot (pixels)
19
20 . # processing variables
21   sortkey=fldr     # sort key (usually fldr or cdp)
22   firsts=800       # first sort (fldr or cdp) value
23    lasts=975       # last sort (fldr or cdp) value
24   increment=5      # sort key increment
25   tracekey=tracf   # trace label key
26
27   #==============================================
```

One of the important reasons we examine the shot gathers is to determine whether we have bad hydrophones (or for land data, geophones). A bad "phone" is easy to find – usually the entire trace has extremely high amplitudes or zero amplitudes (a "dead" trace). An extremely high amplitude (noisy) trace should usually have its amplitudes replaced by zeros ("killed"); a dead trace usually does not cause processing problems.

Figure 12.2 shows shot gather (*fldr*) 930. If you zoom the figure, you can see that traces (*tracf*) 61, 62, and 143 seem to have anomalously high amplitudes. We confirmed this with program **sudumptrace**. (We describe program **sudumptrace** in Appendix C.) Below is the command we entered to use **sudumptrace** and the last ten lines of the screen output. We used **suwind** twice to extract four traces from shot (*fldr*) 930.

```
suwind < Tdata/Tshot5.su key=fldr min=930 max=930 | suwind key=tracf min=60
max=63 | sudumptrace
    1242     4.968    -1.7802e-01   6.6929e+03      6.6929e+03     -1.0794e+00
    1243     4.972    -9.1427e-02   6.6929e+03      6.6929e+03     -8.5523e-01
    1244     4.976    -4.6168e-01   6.6929e+03      6.6929e+03     -5.7248e-01
    1245     4.980    -1.3838e-01   6.6929e+03      6.6929e+03     -6.9779e-01
    1246     4.984     7.6338e-01   6.6929e+03      6.6929e+03     -1.3278e+00
    1247     4.988     1.2673e+00   6.6929e+03      6.6929e+03     -1.5405e+00
    1248     4.992     7.0116e-01   6.6929e+03      6.6929e+03     -1.3003e+00
    1249     4.996    -8.4290e-02   6.6929e+03      6.6929e+03     -1.1742e+00
    1250     5.000     1.1331e-01   6.6929e+03      6.6929e+03     -1.0646e+00
    1251     5.004     3.0200e-01   6.6929e+03      6.6929e+03     -6.6413e-01
```

The first column is time sample, the second column is time value (seconds). The next four columns are amplitude values for traces 60-63. The amplitude values of traces 61 and 62 are constant at 6692.9 while the amplitudes of traces 60 and 63 vary at lower amplitudes.

Below is the command we entered to use **sudumptrace** on the last four traces in shot (*fldr*) 930. Below that are the last ten lines of the screen output.

```
suwind < Tdata/Tshot5.su key=fldr min=930 max=930 | suwind key=tracf min=141
max=144 | sudumptrace
    1242     4.968    -3.2838e+00  -2.9376e-01   1.7135e+03     -6.8847e-01
    1243     4.972    -3.4147e+00   9.3122e-01   1.7135e+03     -6.8070e-01
    1244     4.976    -2.8242e+00   1.7697e+00   1.7135e+03      4.7962e-01
    1245     4.980    -3.0289e+00   9.1051e-01   1.7135e+03      5.8127e-01
    1246     4.984    -3.4756e+00   4.9018e-01   1.7135e+03     -8.4847e-01
    1247     4.988    -3.3356e+00   1.4016e+00   1.7135e+03     -8.5439e-01
    1248     4.992    -3.4664e+00   1.8355e+00   1.7135e+03      5.6005e-01
    1249     4.996    -3.2585e+00   8.9739e-01   1.7135e+03      3.7785e-01
    1250     5.000    -3.1534e+00   1.8295e-02   1.7135e+03     -4.5137e-01
    1251     5.004    -3.5228e+00   4.7177e-02   1.7135e+03      6.9378e-01
```

The first column is time sample, the second column is time value (seconds). The next four columns are amplitude values for traces 141-144. The amplitude values of trace 143 are constant at 1713.5 while the amplitudes of the other traces vary at lower amplitudes.

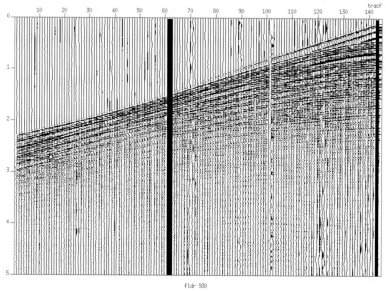

Figure 12.2: *Shot gather (fldr) 930 from file Tshot5.su.*

As we examine the shot gathers with *iview*, we find that every gather has these same three bad *tracf* traces; that is, we see the output of the same three bad hydrophones. In the next section, we will zero ("kill") the amplitude values of traces with *tracf* values 61, 62, and 143.

Note: It is rare that traces are actually removed from a file. By keeping every trace in our Taiwan shot gathers, every *fldr* continues to have exactly 144 traces. This regularity can be useful in (is required by) some processing programs.

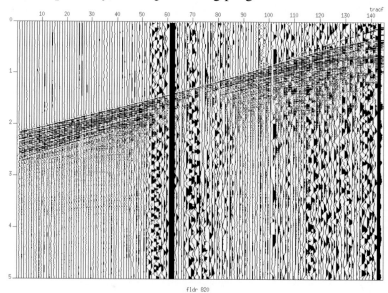

Figure 12.3: *Shot gather (fldr) 820 from file Tshot5.su.*

12.4   Kill Traces

The following lines are from the **sukill** selfdoc:

```
SUKILL - zero out traces

sukill <stdin >stdout min= count=1

Required parameters
        min=                 first trace to kill (one-based)

Optional parameters
        count= 1             number of traces to kill
```

Program **sukill** does not use a key; it finds traces by counting them. For our first shot gather, we have to use **sukill** twice:

    min=61 count=2
    min=143 count=1

For our second shot gather, we have to use **sukill** twice:

    min=205 count=2
    min=287 count=1

We need 2 x 176 uses of **sukill**. Below is script *killer.sh* that generates text for **sukill**.

```
1   #! /bin/sh
2   # File: killer.sh
3   #        Generate sukill values for Taiwan shots
4
5   # Set messages on
6   ##set -x
7
8   outtext=killer.txt
9   rm -f $outtext  # remove earlier trials
10
11  increment=144    # number of traces per shot gather
12  firsts=800       # first fldr value
13  lasts=975        #  last fldr value
14
15  bad2=61          # bad traces count 61,62
16  bad1=143         # bad trace count 143
17
18  #------------------------------------------------
19
20  k=0                                  # fldr counter
21  j=`expr $lasts - $firsts + 1`
22  i=1                                  # loop counter
23  while [ $i -le $j ]
24  do
25
26    two=`expr $k + $bad2`
27    one=`expr $k + $bad1`
28    echo "sukill min=$two count=2 | sukill min=$one count=1 |" >> $outtext
29
30    k=`expr $k + $increment`          # fldr counter
31    i=`expr $i + 1`                   # loop counter
32
33  done
34
35  #------------------------------------------------
36
37  # Exit politely from shell
38  exit
39
```

Below is script *killer2.sh* that zeros the problem traces. Lines 11-186 were generated by *killer.sh* (except that on line 11 we added "< $indata" and on line 186 we replaced the pipe (|) with "> $outdata".

```
 1   #! /bin/sh
 2   # File: killer2.sh
 3   #        kill regularly appearing bad traces
 4
 5   # Set messages on
 6   set -x
 7
 8    indata=Tdata/Tshot5.su
 9   outdata=Tdata/Tshot5k.su
10
11   sukill < $indata min=61 count=2 | sukill min=143 count=1 |
12   sukill min=205 count=2 | sukill min=287 count=1 |
13   sukill min=349 count=2 | sukill min=431 count=1 |
14   sukill min=493 count=2 | sukill min=575 count=1 |
15   sukill min=637 count=2 | sukill min=719 count=1 |
16   sukill min=781 count=2 | sukill min=863 count=1 |
17   sukill min=925 count=2 | sukill min=1007 count=1 |
18   sukill min=1069 count=2 | sukill min=1151 count=1 |
19   sukill min=1213 count=2 | sukill min=1295 count=1 |
20   sukill min=1357 count=2 | sukill min=1439 count=1 |
21   sukill min=1501 count=2 | sukill min=1583 count=1 |
22   sukill min=1645 count=2 | sukill min=1727 count=1 |
23   sukill min=1789 count=2 | sukill min=1871 count=1 |
24   sukill min=1933 count=2 | sukill min=2015 count=1 |
25   sukill min=2077 count=2 | sukill min=2159 count=1 |
26   sukill min=2221 count=2 | sukill min=2303 count=1 |
27   sukill min=2365 count=2 | sukill min=2447 count=1 |
28   sukill min=2509 count=2 | sukill min=2591 count=1 |
29   sukill min=2653 count=2 | sukill min=2735 count=1 |
30   sukill min=2797 count=2 | sukill min=2879 count=1 |
31   sukill min=2941 count=2 | sukill min=3023 count=1 |
32   sukill min=3085 count=2 | sukill min=3167 count=1 |
33   sukill min=3229 count=2 | sukill min=3311 count=1 |
34   sukill min=3373 count=2 | sukill min=3455 count=1 |
35   sukill min=3517 count=2 | sukill min=3599 count=1 |
36   sukill min=3661 count=2 | sukill min=3743 count=1 |
37   sukill min=3805 count=2 | sukill min=3887 count=1 |
38   sukill min=3949 count=2 | sukill min=4031 count=1 |
39   sukill min=4093 count=2 | sukill min=4175 count=1 |
40   sukill min=4237 count=2 | sukill min=4319 count=1 |
41   sukill min=4381 count=2 | sukill min=4463 count=1 |
42   sukill min=4525 count=2 | sukill min=4607 count=1 |
43   sukill min=4669 count=2 | sukill min=4751 count=1 |
44   sukill min=4813 count=2 | sukill min=4895 count=1 |
45   sukill min=4957 count=2 | sukill min=5039 count=1 |
46   sukill min=5101 count=2 | sukill min=5183 count=1 |
47   sukill min=5245 count=2 | sukill min=5327 count=1 |
48   sukill min=5389 count=2 | sukill min=5471 count=1 |
49   sukill min=5533 count=2 | sukill min=5615 count=1 |
50   sukill min=5677 count=2 | sukill min=5759 count=1 |
51   sukill min=5821 count=2 | sukill min=5903 count=1 |
52   sukill min=5965 count=2 | sukill min=6047 count=1 |
53   sukill min=6109 count=2 | sukill min=6191 count=1 |
54   sukill min=6253 count=2 | sukill min=6335 count=1 |
55   sukill min=6397 count=2 | sukill min=6479 count=1 |
56   sukill min=6541 count=2 | sukill min=6623 count=1 |
57   sukill min=6685 count=2 | sukill min=6767 count=1 |
58   sukill min=6829 count=2 | sukill min=6911 count=1 |
59   sukill min=6973 count=2 | sukill min=7055 count=1 |
```

```
 60   sukill min=7117  count=2 |  sukill min=7199  count=1 |
 61   sukill min=7261  count=2 |  sukill min=7343  count=1 |
 62   sukill min=7405  count=2 |  sukill min=7487  count=1 |
 63   sukill min=7549  count=2 |  sukill min=7631  count=1 |
 64   sukill min=7693  count=2 |  sukill min=7775  count=1 |
 65   sukill min=7837  count=2 |  sukill min=7919  count=1 |
 66   sukill min=7981  count=2 |  sukill min=8063  count=1 |
 67   sukill min=8125  count=2 |  sukill min=8207  count=1 |
 68   sukill min=8269  count=2 |  sukill min=8351  count=1 |
 69   sukill min=8413  count=2 |  sukill min=8495  count=1 |
 70   sukill min=8557  count=2 |  sukill min=8639  count=1 |
 71   sukill min=8701  count=2 |  sukill min=8783  count=1 |
 72   sukill min=8845  count=2 |  sukill min=8927  count=1 |
 73   sukill min=8989  count=2 |  sukill min=9071  count=1 |
 74   sukill min=9133  count=2 |  sukill min=9215  count=1 |
 75   sukill min=9277  count=2 |  sukill min=9359  count=1 |
 76   sukill min=9421  count=2 |  sukill min=9503  count=1 |
 77   sukill min=9565  count=2 |  sukill min=9647  count=1 |
 78   sukill min=9709  count=2 |  sukill min=9791  count=1 |
 79   sukill min=9853  count=2 |  sukill min=9935  count=1 |
 80   sukill min=9997  count=2 |  sukill min=10079 count=1 |
 81   sukill min=10141 count=2 |  sukill min=10223 count=1 |
 82   sukill min=10285 count=2 |  sukill min=10367 count=1 |
 83   sukill min=10429 count=2 |  sukill min=10511 count=1 |
 84   sukill min=10573 count=2 |  sukill min=10655 count=1 |
 85   sukill min=10717 count=2 |  sukill min=10799 count=1 |
 86   sukill min=10861 count=2 |  sukill min=10943 count=1 |
 87   sukill min=11005 count=2 |  sukill min=11087 count=1 |
 88   sukill min=11149 count=2 |  sukill min=11231 count=1 |
 89   sukill min=11293 count=2 |  sukill min=11375 count=1 |
 90   sukill min=11437 count=2 |  sukill min=11519 count=1 |
 91   sukill min=11581 count=2 |  sukill min=11663 count=1 |
 92   sukill min=11725 count=2 |  sukill min=11807 count=1 |
 93   sukill min=11869 count=2 |  sukill min=11951 count=1 |
 94   sukill min=12013 count=2 |  sukill min=12095 count=1 |
 95   sukill min=12157 count=2 |  sukill min=12239 count=1 |
 96   sukill min=12301 count=2 |  sukill min=12383 count=1 |
 97   sukill min=12445 count=2 |  sukill min=12527 count=1 |
 98   sukill min=12589 count=2 |  sukill min=12671 count=1 |
 99   sukill min=12733 count=2 |  sukill min=12815 count=1 |
100   sukill min=12877 count=2 |  sukill min=12959 count=1 |
101   sukill min=13021 count=2 |  sukill min=13103 count=1 |
102   sukill min=13165 count=2 |  sukill min=13247 count=1 |
103   sukill min=13309 count=2 |  sukill min=13391 count=1 |
104   sukill min=13453 count=2 |  sukill min=13535 count=1 |
105   sukill min=13597 count=2 |  sukill min=13679 count=1 |
106   sukill min=13741 count=2 |  sukill min=13823 count=1 |
107   sukill min=13885 count=2 |  sukill min=13967 count=1 |
108   sukill min=14029 count=2 |  sukill min=14111 count=1 |
109   sukill min=14173 count=2 |  sukill min=14255 count=1 |
110   sukill min=14317 count=2 |  sukill min=14399 count=1 |
111   sukill min=14461 count=2 |  sukill min=14543 count=1 |
112   sukill min=14605 count=2 |  sukill min=14687 count=1 |
113   sukill min=14749 count=2 |  sukill min=14831 count=1 |
114   sukill min=14893 count=2 |  sukill min=14975 count=1 |
115   sukill min=15037 count=2 |  sukill min=15119 count=1 |
116   sukill min=15181 count=2 |  sukill min=15263 count=1 |
117   sukill min=15325 count=2 |  sukill min=15407 count=1 |
118   sukill min=15469 count=2 |  sukill min=15551 count=1 |
119   sukill min=15613 count=2 |  sukill min=15695 count=1 |
120   sukill min=15757 count=2 |  sukill min=15839 count=1 |
121   sukill min=15901 count=2 |  sukill min=15983 count=1 |
122   sukill min=16045 count=2 |  sukill min=16127 count=1 |
```

```
123   sukill min=16189 count=2 |  sukill min=16271 count=1 |
124   sukill min=16333 count=2 |  sukill min=16415 count=1 |
125   sukill min=16477 count=2 |  sukill min=16559 count=1 |
126   sukill min=16621 count=2 |  sukill min=16703 count=1 |
127   sukill min=16765 count=2 |  sukill min=16847 count=1 |
128   sukill min=16909 count=2 |  sukill min=16991 count=1 |
129   sukill min=17053 count=2 |  sukill min=17135 count=1 |
130   sukill min=17197 count=2 |  sukill min=17279 count=1 |
131   sukill min=17341 count=2 |  sukill min=17423 count=1 |
132   sukill min=17485 count=2 |  sukill min=17567 count=1 |
133   sukill min=17629 count=2 |  sukill min=17711 count=1 |
134   sukill min=17773 count=2 |  sukill min=17855 count=1 |
135   sukill min=17917 count=2 |  sukill min=17999 count=1 |
136   sukill min=18061 count=2 |  sukill min=18143 count=1 |
137   sukill min=18205 count=2 |  sukill min=18287 count=1 |
138   sukill min=18349 count=2 |  sukill min=18431 count=1 |
139   sukill min=18493 count=2 |  sukill min=18575 count=1 |
140   sukill min=18637 count=2 |  sukill min=18719 count=1 |
141   sukill min=18781 count=2 |  sukill min=18863 count=1 |
142   sukill min=18925 count=2 |  sukill min=19007 count=1 |
143   sukill min=19069 count=2 |  sukill min=19151 count=1 |
144   sukill min=19213 count=2 |  sukill min=19295 count=1 |
145   sukill min=19357 count=2 |  sukill min=19439 count=1 |
146   sukill min=19501 count=2 |  sukill min=19583 count=1 |
147   sukill min=19645 count=2 |  sukill min=19727 count=1 |
148   sukill min=19789 count=2 |  sukill min=19871 count=1 |
149   sukill min=19933 count=2 |  sukill min=20015 count=1 |
150   sukill min=20077 count=2 |  sukill min=20159 count=1 |
151   sukill min=20221 count=2 |  sukill min=20303 count=1 |
152   sukill min=20365 count=2 |  sukill min=20447 count=1 |
153   sukill min=20509 count=2 |  sukill min=20591 count=1 |
154   sukill min=20653 count=2 |  sukill min=20735 count=1 |
155   sukill min=20797 count=2 |  sukill min=20879 count=1 |
156   sukill min=20941 count=2 |  sukill min=21023 count=1 |
157   sukill min=21085 count=2 |  sukill min=21167 count=1 |
158   sukill min=21229 count=2 |  sukill min=21311 count=1 |
159   sukill min=21373 count=2 |  sukill min=21455 count=1 |
160   sukill min=21517 count=2 |  sukill min=21599 count=1 |
161   sukill min=21661 count=2 |  sukill min=21743 count=1 |
162   sukill min=21805 count=2 |  sukill min=21887 count=1 |
163   sukill min=21949 count=2 |  sukill min=22031 count=1 |
164   sukill min=22093 count=2 |  sukill min=22175 count=1 |
165   sukill min=22237 count=2 |  sukill min=22319 count=1 |
166   sukill min=22381 count=2 |  sukill min=22463 count=1 |
167   sukill min=22525 count=2 |  sukill min=22607 count=1 |
168   sukill min=22669 count=2 |  sukill min=22751 count=1 |
169   sukill min=22813 count=2 |  sukill min=22895 count=1 |
170   sukill min=22957 count=2 |  sukill min=23039 count=1 |
171   sukill min=23101 count=2 |  sukill min=23183 count=1 |
172   sukill min=23245 count=2 |  sukill min=23327 count=1 |
173   sukill min=23389 count=2 |  sukill min=23471 count=1 |
174   sukill min=23533 count=2 |  sukill min=23615 count=1 |
175   sukill min=23677 count=2 |  sukill min=23759 count=1 |
176   sukill min=23821 count=2 |  sukill min=23903 count=1 |
177   sukill min=23965 count=2 |  sukill min=24047 count=1 |
178   sukill min=24109 count=2 |  sukill min=24191 count=1 |
179   sukill min=24253 count=2 |  sukill min=24335 count=1 |
180   sukill min=24397 count=2 |  sukill min=24479 count=1 |
181   sukill min=24541 count=2 |  sukill min=24623 count=1 |
182   sukill min=24685 count=2 |  sukill min=24767 count=1 |
183   sukill min=24829 count=2 |  sukill min=24911 count=1 |
184   sukill min=24973 count=2 |  sukill min=25055 count=1 |
185   sukill min=25117 count=2 |  sukill min=25199 count=1 |
```

```
186   sukill min=25261 count=2 | sukill min=25343 count=1 > $outdata
187
188   # Exit politely from shell
189   exit
190
```

Figure 12.4 shows shot gathers 930 and 820 after zeroing *tracr* traces 61, 62, and 143. The ellipses on shot gather 820 show where 1-D frequency analysis is done in the next section.

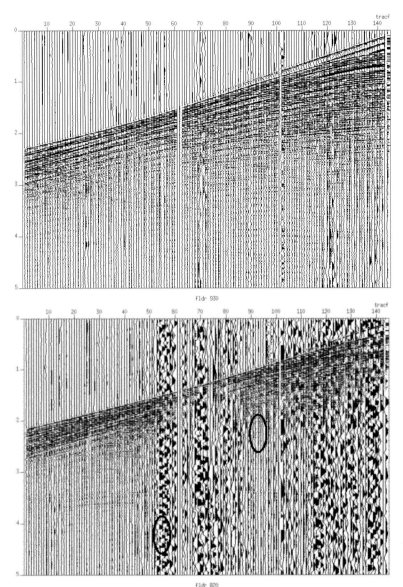

Figure 12.4: *Shot (fldr) gathers 930 (top) and 820 (bottom) from file Tshot5k.su.*

Note: In the Taiwan shot-order data, these bad traces have regular position. In CMP-order data, these bad traces have no regular position. If we had not examined the gathers before sorting them to CMP order, we would not have recognized the regularity of these traces (that they were caused by particular bad hydrophones) and we would have to use a different way to identify these traces. One other way to identify these traces is by their regular high amplitudes.

12.5 One-dimensional Frequency Analysis

In Section 10.4.1, we used script *fxdisp* to see the frequencies of a gather. The frequency display of *fxdisp* is the transform of each trace. Here, we present script *tf.sh* that plots frequencies of a single trace or part of a single trace. Our examples use a half-second of data from traces windowed from shot gather 820, shown by the ellipses in the bottom shot gather of Figure 12.4.

In Figure 12.5, the left side example is from a low frequency noise area: *tracf*=55, 4-4.5 seconds.

```
suwind < Tdata/Tshot5k.su key=fldr min=820 max=820 | suwind key=tracf min=55
    max=55 tmin=4 tmax=4.5 > T820f55t4.su
```

We created the left side plots with the following command:

```
tf.sh  T820f55t4.su
```

The right side example in Figure 12.5 is from a good signal area: *tracf*=93, 2-2.5 seconds.

```
suwind < Tdata/Tshot5k.su key=fldr min=820 max=820 | suwind key=tracf min=93
    max=93 tmin=2 tmax=2.5 > T820f93t2.su
```

We created the right side plots with the following command:

```
tf.sh  T820f93t2.su
```

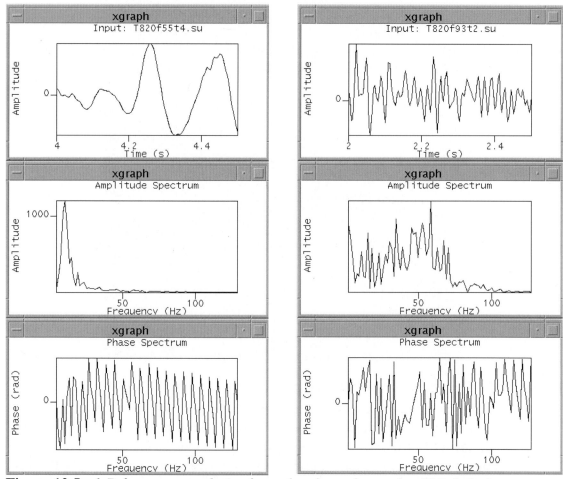

Figure 12.5: *1-D frequency analysis of wavelets from shot gather 820. Left: low frequency noise wavelet. Right: good high frequency wavelet.*

Script *tf.sh* makes three plots. The top plot is the input time series, the middle plot is the frequency transform of the time series, and the bottom plot is the phase spectrum of the time series. Below is script *tf.sh*.

```
 1  #! /bin/sh
 2  # File: tf.sh
 3  #         Time to frequency transform of a trace
 4  #   Output = 3 plots: time, freq amplitude, phase spectrum
 5  #       Use: tf.sh [input.su]
 6  # Example: tf.sh wave1.su
 7
 8  # Set messages on
 9  set -x
10
11  suxgraph < $1 -geometry 400x200+10+10 -bg white \
12          title="Input: $1" \
13          label1="Time (s)" label2="Amplitude" \
14          style=normal linecolor=0 &
15
16  ##suxwigb < $1 wt=1 va=2 style=normal labelcolor=black \
17  ##          'wbox=400 hbox=200 xbox=420 ybox=10 \
18  ##          title="Input: $1" titlecolor=black \
19  ##          label1="Time (s)" label2=" Amplitude" &
20
21  sufft < $1 | suamp mode=amp |
22  suxgraph -geometry 400x200+10+245 -bg white \
23          title="Amplitude Spectrum" \
24          label1="Frequency (Hz)" label2="Amplitude" \
25          style=normal linecolor=0 &
26
27  sufft < $1 | suamp mode=phase |
28  suxgraph -geometry 400x200+10+480 -bg white \
29          title="Phase Spectrum" \
30          label1="Frequency (Hz)" label2="Phase (rad)" \
31          style=normal linecolor=0 &
32
33  # "press return key to ..."
34  pause remove the plots
35
36  zap xgraph
37  ##zap xwigb
38
39  # Exit politely from shell
40  exit
41
```

- Line 6: As the example shows, the input file must be supplied when the script is used. The input file is variable *$1*.

- Lines 11-14 make the time series plot with **suxgraph**.

- Line 21: **sufft** transforms the data and **suamp** mode=amp outputs the amplitude spectrum of the information from **sufft**.

- Lines 22-25 plot the amplitude spectrum with **suxgraph**.

- Line 27: **sufft** transforms the data and **suamp** mode=phase outputs the phase spectrum of the information from **sufft**.

- Lines 28-31 plot the phase spectrum with **suxgraph**.

- Lines 16-19 make an alternative time series plot using **suxwigb**. These lines can be uncommented for use.

Note: If you middle click the mouse in an **xwigb** window, mouse location information is printed in the upper left of the window. An **xgraph** window does not have this feature.

- After the plots are made, Line 34 causes the following line to be printed to the screen: `press return key to remove the plots`

Script *tf.sh* is a display script; no data are output.

13. Taiwan: Gain-Filter, Filter-Gain

13.1 Introduction

We now have 5 seconds of Taiwan data and we zeroed the bad hydrophones. In this chapter we will

1. apply spherical divergence correction, then band-pass filter the data.

We will also

2. apply a band-pass filter, then apply spherical divergence correction.

We will compare gathers from these two paths and consider whether one path is better.

13.2 Gain-Filter

13.2.1 Gain – Spherical Divergence Correction

To determine which exponential gain to apply to the line of CMPs, we will apply gain to two shot gathers: 820 and 930.. We window them from the 2-D line:

```
suwind < Tdata/Tshot5k.su key=fldr min=820 max=820 > T5k820.su
suwind < Tdata/Tshot5k.su key=fldr min=930 max=930 > T5k930.su
```

We will determine a *tpow* value for **sugain** by using script igain (Section 10.5), option T. Below is the user area of *igain.sh* changed for the Taiwan data.

```
 9   # User-supplied values
10
11   indata=T5k820.su  # Input file
12   myperc=90         # perc value
```

The script creates three displays:

*   a wiggle plot of the file,
*   a decibel (dB) ximage plot of the user-selected trace(s), and
*   an amplitude graph of the user-selected trace(s).

Remember:

*   The script applies gain to the entire file, but only the user-selected trace or traces are used to create the second and third displays.

*   The user is asked to supply the name of a key (*offset*, *tracr*, etc.) and corresponding key minimum and key maximum values for the second and third displays.

We selected key *tracf* and supplied *tracf* values 92 93 for both shot gathers, based on our displays in Chapter 12.

Table 13.1: dB ranges of Figures 13.1 and 13.2

fldr	tracf		No gain	tpow = 1.8
820	92 93	maximum dB	0	0
		minimum dB	-79.4	-110.6
930	92 93	maximum dB	0	0
		minimum dB	-82.53	-123.9

After several trials, we decide to use a spherical divergence correction of 1.8 (Figures 13.1 and 13.2).

We apply spherical divergence correction with the command below.

```
sugain  <  Tdata/Tshot5k.su  tpow=1.8  >  Tdata/Tshot5kg.su
```

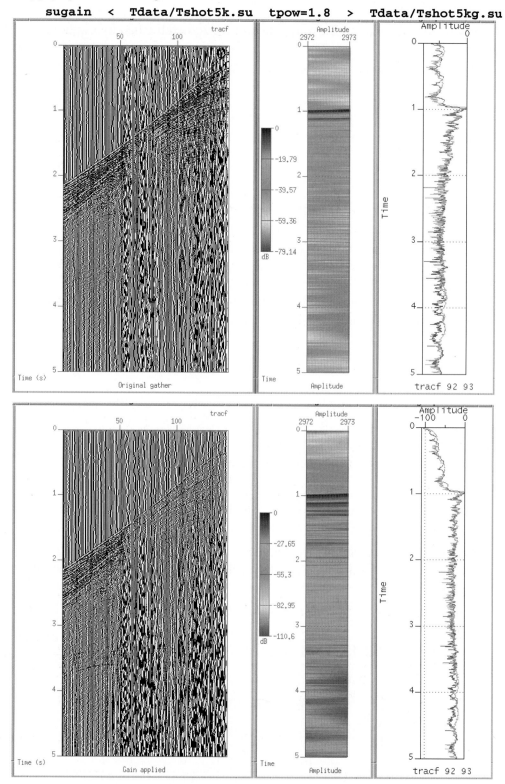

Figure 13.1: *Top: Shot 820 without gain. Bottom: Shot 820 with gain, tpow=1.8.*

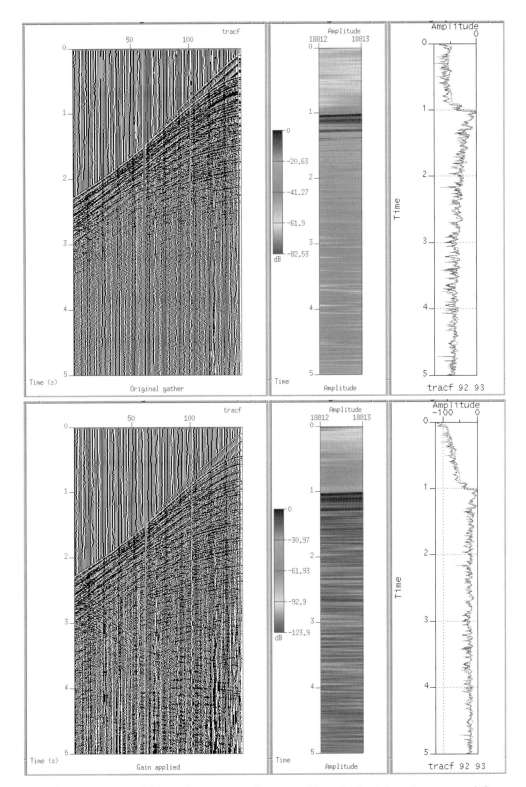

Figure 13.2: *Top: Shot 930 without gain. Bottom: Shot 930 with gain, tpow=1.8.*

13.2.2 Band-pass Filter

To select frequencies for band-pass filtering, we window two shot gathers from the gained 2-D line and examine them with script *fxdisp*.

```
suwind < Tdata/Tshot5kg.su key=fldr min=820 max=820 > T5kg820.su
suwind < Tdata/Tshot5kg.su key=fldr min=930 max=930 > T5kg930.su
```

The *fxdisp* dialog asks for three values:

- the input file name,
- a *perc* value for the wiggle plot of the input file, and
- a key (trace header) to label the x-axis of the wiggle plot.

The script creates two displays: a wiggle plot of the input traces and a corresponding frequency display. The script transforms each input seismic trace to a frequency trace.

The script repeats these three questions and two displays until you quit the script. No data are saved after processing.

If you make a mistake while typing, use the Delete key, not the Back Space key.

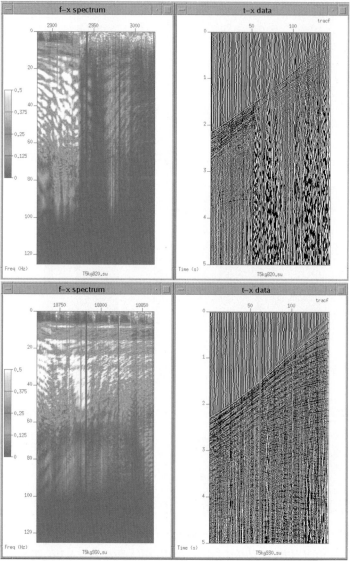

Figure 13.3: *Plots of f-x and t-x after gain. Top: Shot 820. Bottom: Shot 930.*

The top of Figure 13.3 shows that the low frequency noise overwhelms the reflections.

Note: The x-axis annotation of the frequency plot (an **xwigb** window) shows *tracr* values, the first default key. See Section 7.7.

Based on Figure 13.3, we will use the following command to filter the gained data:

```
sufilter < Tdata/Tshot5kg.su f=18,23,55,60 amps=0,1,1,0 > Tdata/Tshot5kgf.su
```

13.2.3 Gain-Filter Result

Figure 13.4 shows our two test shot gathers after spherical divergence correction followed by our band-pass filter.

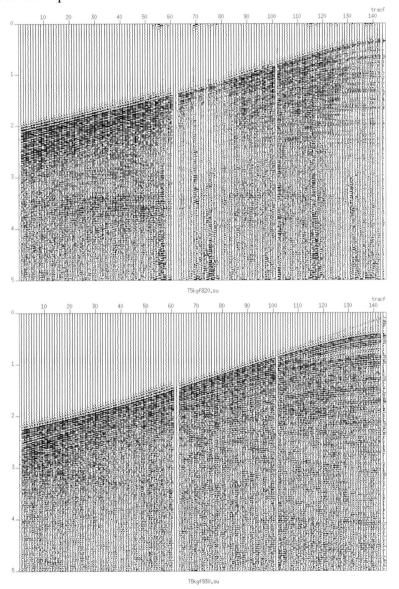

Figure 13.4: *After gain, then band-pass filter. Top: Shot 820. Bottom: Shot 930.*

Let's continue our experiments by reversing the process order.

13.3 Filter-Gain

13.3.1 Band-pass Filter

To select frequencies for band-pass filtering, we examine the two shot gathers we windowed from the 2-D line at the beginning of Section 13.2: *T5k820.su* and *T5k930.su*.

We examine these files with script *fxdisp*.

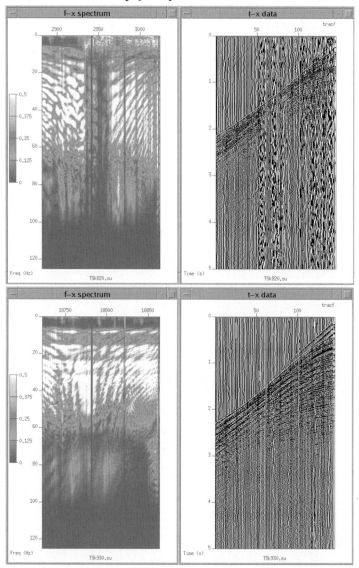

Figure 13.5: *Plots of f-x and t-x. Top: Shot 820. Bottom: Shot 930.*

Compare Figure 13.5, no gain, to Figure 13.3, *fxdisp* after gain. Without gain, the low frequency noise on the near-offset traces of gather 820 does not overwhelm the reflections.

After discussion, we decide to use the same **sufilter** values we used in Section 13.3.

```
sufilter < Tdata/Tshot5k.su f=18,23,55,60 amps=0,1,1,0 > Tdata/Tshot5kf.su
```

13.3.2 Gain – Spherical Divergence Correction

We will apply exponential gain to two shot gathers windowed from the filtered data:

```
suwind < Tdata/Tshot5kf.su key=fldr min=820 max=820 > T5kf820.su
suwind < Tdata/Tshot5kf.su key=fldr min=930 max=930 > T5kf930.su
```

We will determine a *tpow* value for **sugain** by using script *igain*, option T. Again, we will use key *tracf* and *tracf* values 92 93.

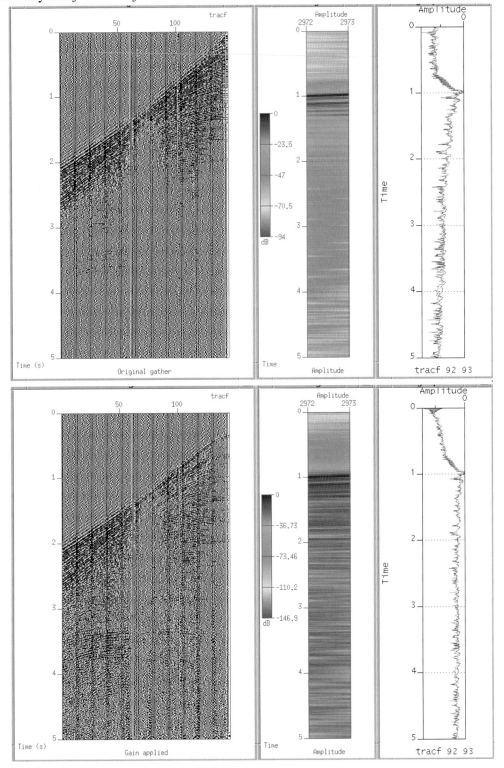

Figure 13.6: *Top: Shot 820 without gain. Bottom: Shot 820 with gain, tpow=2.0.*

This time, we think a *tpow* value of 2.0 nicely balances the data (Figures 13.6, 13.7).

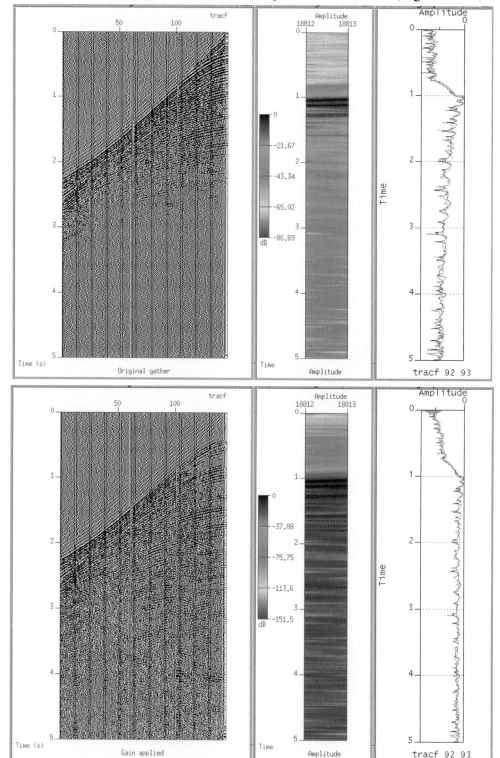

Figure 13.7: *Top: Shot 930 without gain. Bottom: Shot 930 with gain, tpow=2.0.*
Table 13.2 shows the amplitude range increase of spherical divergence correction.
We apply spherical divergence correction with the command below.

```
sugain  <  Tdata/Tshot5kf.su  tpow=2.0  >  Tdata/Tshot5kfg.su
```

Table 13.2: dB ranges of Figures 13.5 and 13.6

fldr	tracf		No gain	tpow = 2.0
820	92 93	maximum dB	0	0
		minimum dB	-94	-146.9
930	92 93	maximum dB	0	0
		minimum dB	-86.59	-151.5

13.3.3 Filter-Gain Result

Figure 13.8 shows our two test shot gathers after band-pass filtering followed by spherical divergence correction.

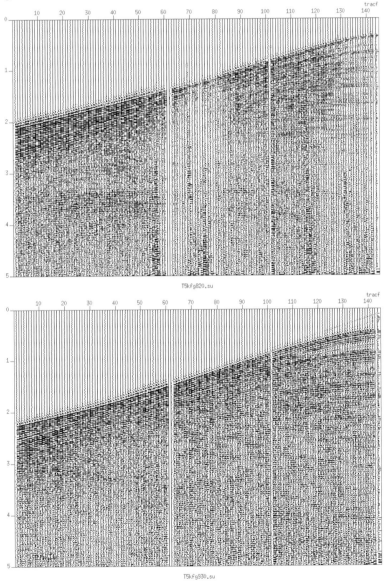

Figure 13.8: *After band-pass filter, then gain. Top: Shot 820. Bottom: Shot 930.*

13.4 Discussion

When we compare Figures 13.4 and 13.8, we see minor differences, but not major differences. Recall that we used different gain values in the two tests. While we used the same frequency pass-band for both tests, we did consider passing slightly more lower frequencies in the second test (Section 13.3.1).

Further analysis can be done using **sudiff** to get a difference file. For example:

```
sudiff   T5kgf820.su   T5kfg820.su   >   Tdiff820.su
sudiff   T5kgf930.su   T5kfg930.su   >   Tdiff930.su
```

To see the documentation for **sudiff**, enter "sudiff" or "sudoc sudiff".

Also, you can examine trace amplitudes by using **suascii** or **sudumptrace**. Program **sudumptrace** is described in Appendix C.

14. f-k Filter and Deconvolution

14.1 Introduction

Real data have noise of various kinds. Some signal enhancement programs are used prestack (before stacking) and some are used post-stack (after stacking CMPs). In the previous chapter, we used gain and band-pass filtering on prestack data. We only used one option out of many in **sugain**, but you are free to test other **sugain** options. In this chapter, we present two popular signal enhancement tools. These tools could prove useful in processing the Taiwan data set; however, we were unable to improve that data set using them. (Perhaps you can find a way!). To demonstrate their capability, we use them on Oz files (shot gathers) that we get from the Colorado School of Mines Center for Wave Phenomena web site (Section 3.3).

The first tool is a frequency-wavenumber (f-k) filter, a two-dimensional filter. We see how to use an f-k filter through the **sudipfilt** program. For a brief description of the f-k domain, see Sheriff (2002): f-k domain.

The second tool is deconvolution. We implement a deconvolution (decon) script that uses **suacor**, an autocorrelation program, and **supef**, a prediction-error filter.

14.2 f-k Filter -- Absolute Units

In Section 10.4.1, we used the one-dimensional transform display script *fxdisp*. Script *fxdisp* transforms each trace from a time series to a frequency series. Here, we use interactive script *ifk* to apply a two-dimensional frequency filter. The f-k filter slopes can be specified in relative units of time over trace spacing (for example, seconds/meter) or in absolute units of time samples per trace. In this section, we specify the f-k filter slopes in absolute units: we measure the vertical axis as "number of samples" and we measure the horizontal axis as "number of traces."

14.2.1 Use the f-k Filter Script

Script *ifk* interactively lets us first view the f-k spectrum of a gather, then specify the slopes of the pass/reject filter. In this example, the input file is Oz gather 8. We get Oz gather 8, a split-spread gather, from the Colorado School of Mines web site. Below is the **surange** output of *oz08.su*.

```
        surange < oz08.su
96 traces:
 tracl=(1,96)  tracr=(1,96)  fldr=10008 tracf=(1,96)  cdp=(8,103)
 cdpt=1 trid=1 nvs=1 nhs=1 duse=1
 scalel=1 scalco=1 counit=1 delrt=4 muts=4
 ns=1350 dt=4000
```

However, we want to use only one side of the split-spread gather. Also, *oz08.su* has 5.4 seconds; we will window that to the first 3 seconds.

```
        suwind < oz08.su key=tracl min=26 max=96 tmax=3 > oz08w.su
```

Below is the **surange** output of *oz08w.su*.

```
        surange  <  oz08w.su
71 traces:
 tracl=(26,96)  tracr=(26,96)  fldr=10008 tracf=(26,96)  cdp=(33,103)
 cdpt=1 trid=1 nvs=1 nhs=1 duse=1
```

```
scalel=1 scalco=1 counit=1 delrt=4 muts=4
ns=750 dt=4000
```

Below is the screen dialog with our input highlighted. Line numbers are added for discussion. We input values on lines 14, 21, and 27. Note that our slope values, supplied on dialog line 21, are 4,5,6,7 (absolute units).

```
 1     -------------------------------------------------------
 2                      f-k Filter Test
 3                      ---------------
 4     From your slope values, two filters are created --
 5          a pass filter and a reject filter.
 6     When you exit, the following files are output:
 7                 Slopes are in   ==>  fk.txt
 8            Passed data are in   ==>  fkpass.su
 9          Rejected data are in   ==>  fkrejj.su
10     -------------------------------------------------------
11
12     Supply gain power value for t^(power)
13     For no gain, supply 0
14     2
15
16     Supply filter slopes.
17     Input: s1,s2,s3,s4     where s1 < s2 < s3 < s4
18        Example:   3.0,3.5,4.0,4.5
19           or:    -4.5,-4.0,-3.5,-3.0
20     Use commas. Do not use spaces.
21     4,5,6,7
22
```

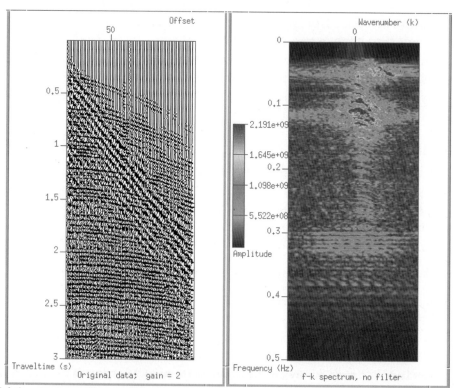

Figure 14.1: *Input gather oz08w.su. Left: Seismic after gain (tpow=2.0). Right: f-k plot of seismic after gain.*

```
23                 Slopes: 4,5,6,7
24
25     Enter 1 for more f-k filter testing
```

```
26  Enter 2 to EXIT
27  2
28
29         Slopes are in  ==>  fk.txt
30      Passed data are in  ==>  fkpass.su
31  Rejected data are in  ==>  fkrejj.su
32
33  press return key to exit
```

The script lets us apply a t^(power) gain to the input seismic data (dialog lines 12-14). After that question is answered, the first displays are made. One shows the input seismic data, the other shows an f-k plot of the seismic data (Figure 14.1).

Using the four user-supplied slope values (dialog lines 16-20), script *ifk* makes two outputs using **sudipfilt**:

- a pass filter

    slopes=s1,s2,s3,s4   amps=0,1,1,0

- a reject filter

    slopes=s1,s2,s3,s4   amps=1,0,0,1

Figure 14.2 shows the four output plots: a t-x plot and an f-k plot after application of the pass filter and a t-x plot and an f-k plot after application of the reject filter. The test can be repeated. When the user exits the script, the seismic results of the last test are output: the pass seismic data (*fkpass.su*) and the reject seismic data (*fkrejj.su*). A log of the processing steps is output as file *fk.txt* (shown below).

```
                f-k Filter Test
                ---------------
sugain < oz08w.su tpow=2
 ==> First test
            Slopes: 4,5,6,7
sudipfilt < oz08w.su dx=1 dt=1 \
        slopes=  amps=
   Passed data are in  ==>  fkpass.su
 Rejected data are in  ==>  fkrejj.su
```

In output file *fk.txt*, no values are output for parameter *amps* because it is left to the user whether to use slopes as a pass filter (amps=0,1,1,0) or as a reject filter (amps=1,0,0,1).

We rename the output files to prevent our files from becoming overwritten by a later run: **Note:** Use a semicolon to separate multiple commands on the same line.

    mv fkpass.su fkpass08abs.su; mv fkrejj.su fkrejj08abs.su

14.2.2 Script ifk.scr

We usually start interactive scripts by running a *.scr* script. Here is *ifk.scr*:

```
1  #! /bin/sh
2  # File: ifk.scr
3  #      Run this script to start script ifk.sh
4
5  xterm -geom 80x17+10+640 -e ifk.sh
6
```

Line 1 invokes the shell. Line 5 opens a dialog window 80 characters wide by 17 characters high. Line 5 also places the dialog window 10 pixels from the left side of the viewing area and 640 pixels from the top of the viewing area. Line 5 also starts *ifk.sh*, the processing script, within that window.

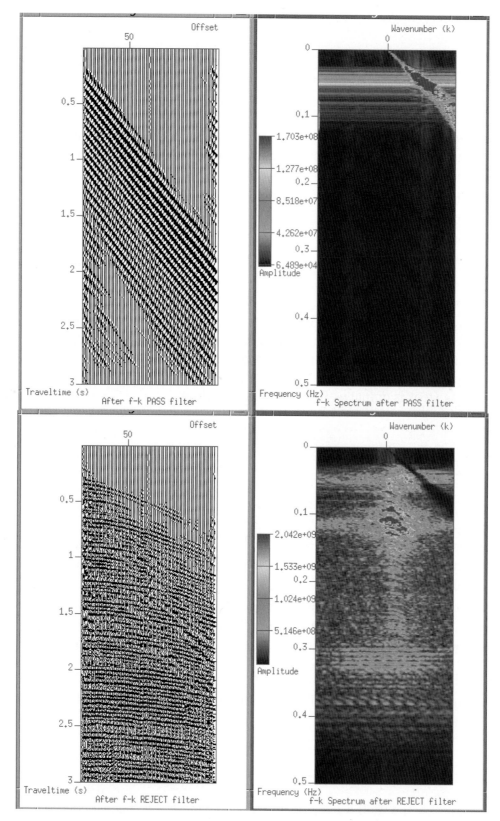

Figure 14.2: *Output of slope values 4,5,6,7. Top: Data passed by filter. Bottom: Data rejected by filter.*

14.2.3 Script ifk.sh

We start interactive scripts by running a *.scr* script. Here is *ifk.sh*.

```
 1  #! /bin/sh
 2  # File: ifk.sh
 3  #       Run script ifk.scr to start this script
 4  #       Interactive f-k processing
 5
 6  # Set messages on
 7  ##set -x
 8
 9  #=================================================
10  # USER AREA -- SUPPLY VALUES
11  #-------------------------------------------------
12
13  # Input seismic file
14  indata=oz08w.su
15
16  # Display choices
17  myperc=99       # perc value for seismic plots
18  plottype=0      # 0 = wiggle plot,  1 = image plot
19
20  # Processing variables [ Instructions below ]
21  dx=1  # trace spacing
22  dt=1  # time sample interval
23
24  #=================================================
25
26  # Instructions:
27  # -----------
28  # sudipfilt is a classic pie slice filter. Each slope radiates
29  #    from the f-k origin. Supply four slope values. A "slope" is
30  #
31  #                         change in t
32  #                         -----------
33  #                         change in x
34  #
35  # You can use relative units: (seconds/meter)
36  #     for example: dt=.004  dx=25
37  #   or absolute units: (time samples / trace)
38  #     for example: dt=1  dx=1
39  #
40  #-------------------------------------------------
41
42  echo " -----------------------------------------------------"
43  echo "                 f-k Filter Test"
44  echo "              --------------"
45  echo "  From your slope values, two filters are created --"
46  echo "       a pass filter and a reject filter."
47  echo "  When you exit, the following files are output:"
48  echo "           Slopes are in   ==>   fk.txt"
49  echo "           Passed data are in   ==>   fkpass.su"
50  echo "           Rejected data are in   ==>   fkrejj.su"
51  echo " -----------------------------------------------------"
52
53  # Remove temporary files
54  rm -f tmp*
55
56  #-------------------------------------------------
57  # Describe temporary files
58  #-------------------------------------------------
59
60  # tmp0 = Binary. A copy of the input seismic file
```

```
61   # tmp1 = Binary. Data that is processed, either tmp0 or tmp2
62   #         tmp1 is a copy of tmp0 if re-processing not wanted
63   #         tmp1 is a copy of tmp2 if re-processing is wanted
64   # tmp2 = Binary. Output of "a=0,1,1,0" (pass) dip filtering
65   # tmp3 = Binary. Output of "a=1,0,0,1" (reject) dip filtering
66   # tmp4 = ASCII record of dialog for output text file
67   # tmp5 = ASCII file to reduce "zap" screen messages
68
69   #--------------------------------------------------
70   # Supply gain value to t^(power)
71   #--------------------------------------------------
72
73   echo " "
74   echo "Supply gain power value for t^(power)"
75   echo "For no gain, supply 0"
76   > /dev/tty
77   read tpow
78
79   # Copy input data to temporary file; apply gain if requested
80   sugain < $indata tpow=$tpow > tmp0
81
82   # Log input file & gain value
83   echo "                    f-k Filter Test" > tmp4
84   echo "                    --------------" >> tmp4
85   echo "sugain < $indata tpow=$tpow" >> tmp4
86
87   #--------------------------------------------------
88   # Plot original gather and spectrum
89   #--------------------------------------------------
90
91   if [ $plottype -eq 0 ] ; then
92     suxwigb < tmp0 xbox=10 ybox=10 wbox=300 hbox=500 \
93               label1=" Time (s)" label2="Offset" \
94               title="Original data;  gain = $tpow" key=offset \
95               perc=$myperc verbose=0 &
96   else
97     suximage < tmp0 xbox=10 ybox=10 wbox=300 hbox=500 \
98               label1=" Time (s)" \
99               title="Original data;  gain = $tpow" \
100              perc=$myperc verbose=0 &
101  fi
102
103  suspecfk < tmp0 dx=$dx dt=$dt |
104  suximage    xbox=320 ybox=10 wbox=300 hbox=500 \
105              label1=" Frequency (Hz)" label2="Wavenumber (k)"\
106              title="f-k spectrum, no filter" \
107              cmap=hsv2 legend=1 units=Amplitude verbose=0 \
108              grid1=dots grid2=dots perc=99 &
109
110  #--------------------------------------------------
111  # f-k filter test
112  #--------------------------------------------------
113
114  new=true  # true = first test
115  ok=false  # false = continue looping
116
117  while [ $ok = false ]
118  do
119
120    rm -f tmp1  # remove earlier copy of file to be filtered
121
122    if [ $new = true ] ; then
123      echo " ==> First test" >> tmp4
```

```
124        cp tmp0 tmp1
125      else
126        echo " "
127        echo "Enter A or a to add an f-k filter"
128        echo "Enter S or s to start over"
129        > /dev/tty
130        read choice1
131
132        case $choice1 in
133          [sS])
134                cp tmp0 tmp1
135                echo " ==> Using original data"
136                echo " ==> Using original data" >> tmp4
137                ;;
138          [aA])
139
140                echo " "
141                echo "Enter P or p to use Passed data"
142                echo "Enter R or r to use Rejected data"
143                > /dev/tty
144                read choice3
145
146                case $choice3 in
147                  [pP])
148                        echo " "
149                        cp tmp2 tmp1
150                        echo " ==> Using passed data"
151                        echo " ==> Using passed data" >> tmp4
152                        ;;
153                  [rR])
154                        echo " "
155                        cp tmp3 tmp1
156                        echo " ==> Using rejected data"
157                        echo " ==> Using rejected data" >> tmp4
158                        ;;
159                esac
160
161                ;;
162        esac
163
164 # Remove earlier test of passed and rejected data
165      rm -f tmp2
166      rm -f tmp3
167
168    fi
169
170  #-----------------------------------------------
171  # Get filter slope values
172  #-----------------------------------------------
173
174    echo " "
175    echo "Supply filter slopes."
176    echo "Input: s1,s2,s3,s4    where s1 < s2 < s3 < s4"
177    echo " Example: 3.0,3.5,4.0,4.5"
178    echo "      or: -4.5,-4.0,-3.5,-3.0"
179    echo "Use commas. Do not use spaces."
180    > /dev/tty
181    read slopes
182    echo " "
183    echo "          Slopes: $slopes"
184    echo "          Slopes: $slopes" >> tmp4
185
186  #-----------------------------------------------
```

```
187  # Apply filters, plot seismic data, plot f-k data
188  #------------------------------------------------
189
190  # Apply pass filter
191    sudipfilt < tmp1 dx=$dx dt=$dt slopes=$slopes amps=0,1,1,0 > tmp2
192
193  # Plot seismic passed data
194    if [ $plottype -eq 0 ] ; then
195      suxwigb < tmp2 xbox=10 ybox=10 wbox=300 hbox=500 \
196                label1=" Time (s)" label2="Offset"\
197                title="After f-k PASS filter" key=offset \
198                perc=$myperc verbose=0 &
199    else
200      suximage < tmp2 xbox=10 ybox=10 wbox=300 hbox=500 \
201                label1=" Time (s)" \
202                title="After f-k PASS filter" \
203                perc=$myperc verbose=0 &
204    fi
205
206  # Plot f-k passed data
207    suspecfk < tmp2 dx=$dx dt=$dt |
208    suximage   xbox=320 ybox=10 wbox=300 hbox=500 \
209                label1=" Frequency (Hz)" label2="Wavenumber (k)" \
210                title="f-k Spectrum after PASS filter" \
211                cmap=hsv2 legend=1 units=Amplitude verbose=0 \
212                grid1=dots grid2=dots perc=99 &
213
214  # Apply reject filter
215    sudipfilt < tmp1 dx=$dx dt=$dt slopes=$slopes amps=1,0,0,1 > tmp3
216
217  # Plot seismic rejected data
218    if [ $plottype -eq 0 ] ; then
219      suxwigb < tmp3 xbox=630 ybox=10 wbox=300 hbox=500 \
220                label1=" Traveltime (s)" label2="Offset" \
221                title="After f-k REJECT filter" key=offset \
222                perc=$myperc verbose=0 &
223    else
224      suximage < tmp3 xbox=630 ybox=10 wbox=300 hbox=500 \
225                label1=" Time (s)" \
226                title="After f-k REJECT filter" \
227                perc=$myperc verbose=0 &
228    fi
229
230  # Plot f-k rejected data
231    suspecfk < tmp3 dx=$dx dt=$dt |
232    suximage   xbox=940 ybox=10 wbox=300 hbox=500 \
233                label1=" Frequency (Hz)" label2="Wavenumber (k)" \
234                title="f-k Spectrum after REJECT filter" \
235                cmap=hsv2 legend=1 units=Amplitude verbose=0 \
236                grid1=dots grid2=dots perc=99 &
237
238  #------------------------------------------------
239  # More f-k or exit
240  #------------------------------------------------
241
242    echo " "
243    echo "Enter 1 for more f-k filter testing"
244    echo "Enter 2 to EXIT"
245    > /dev/tty
246    read choice2
247
248    case $choice2 in
249      1)
```

```
250          ok=false
251          ;;
252       2)
253          cp tmp2 fkpass.su
254          cp tmp3 fkrejj.su
255          echo "sudipfilt < $indata dx=$dx dt=$dt \\" >> tmp4
256          echo "                 slopes=  amps=" >> tmp4
257          echo " "
258          echo "Processing log is in  ==>   fk.txt"
259          echo " Passed data are in  ==>   fkpass.su"
260          echo " Passed data are in  ==>   fkpass.su" >> tmp4
261          echo "Rejected data are in  ==>   fkrejj.su"
262          echo "Rejected data are in  ==>   fkrejj.su" >> tmp4
263          cp tmp4 fk.txt
264          pause exit
265          zap xwigb > tmp5
266          zap ximage > tmp5
267          ok=true
268          ;;
269    esac
270
271    new=false  # true = first test
272
273  done
274
275  #------------------------------------------------
276  # Exit
277  #------------------------------------------------
278
279  # Remove temporary files
280  rm -f tmp*
281
282  # Exit politely from shell
283  exit
284
```

The user-supplied values are on lines 13-22. Lines 21 and 22 are where trace spacing and the time sample interval are set. The next lines, 26-40, discuss these parameters.

Lines 42-51 are written to the screen as soon as the script starts so the user will know what outputs to expect.

Line 54 removes temporary files that might be left from a previously crashed run.

Lines 60-67 document the use of internal files.

Lines 73-77 ask the user if t^(power) gain is desired.

Line 80 copies the seismic data to an internal file while applying gain, or no gain.

Lines 83-85 write the input file name and the gain value to the temporary log files.

Lines 91-101 plot the input seismic data using **suxwigb** or **suximage**.

Line 103 transforms the input seismic data to f-k space. Lines 104-108 plot the 2-D transform of the input seismic data.

Actual filter testing occurs between lines 114 and 273.

Line 114, parameter *new*, is a test for the *if* block of lines 122-168. If this is a first test (the first time through the loop), the input data are copied from *tmp0* to *tmp1* (line 124). If this is not the first test, the user is faced with *choice1*, lines 127-130. If the user chooses to start fresh, the original data in *tmp0* are copied to *tmp1* (line 134). If the user chooses to re-filter data already filtered, another choice (*choice3*) must be made: whether to reprocess the data from the "pass" filter or from the "reject" filter (lines 141-144). After a previous test, the "pass" data are in *tmp2* (line 191) and the "reject" data are in *tmp3* (line 215). The user's choice (*choice3*) determines which file is copied to (overwrites) *tmp1* for the next test (line 149 or line 155). At the bottom of the test loop, line 273, *new=false* insures that, after the first time through the loop, the user will always have to see *choice1*.

If this is not the first time through the loop, lines 165-166 remove the previous "pass" and "reject" data files.

Lines 175-181 explain the input format of the slopes and read the slopes from the user.

Line 191 applies the "pass" filter (amps=0,1,1,0). Lines 194-204 plot the seismic "pass" data. Line 207 transforms the "pass" seismic data to f-k space. Lines 208-212 plot the 2-D transform of the "pass" seismic data.

Line 215 applies the "reject" filter (amps=1,0,0,1). Lines 218-228 plot the seismic "reject" data. Line 231 transforms the "reject" seismic data to f-k space. Lines 232-236 plot the 2-D transform of the "reject" seismic data.

Lines 243-246 offer the user the choice (*choice2*) to re-filter or exit the script.

If the user chooses to re-filter (*choice2=1*), line 250 lets the script pass the user to the top of the loop. If the user chooses to exit (*choice2=2*):

- Lines 253-254 create permanent disk files from the last processed "pass" and "reject" test.

- Lines 255-256 put a template of the **sudipfilt** parameters into the temporary log file.

- Lines 258-262 write information to the screen and the log file.

- Line 263 copies the temporary log file to a permanent disk file.

- Line 264 makes the script pause so the user can read the screen messages.

- Lines 265-266 close the plot windows and re-direct some of the accompanying system messages to a temporary file.

- Line 267 sets a flag (*ok=true*) that ends cycling through the loop (see line 117).

After using *ifk* the first time, line 271 insures that the user is directed to the *else* portion of *if* block lines 122-168.

Line 280 removes all temporary files.

Line 283 exits the script.

14.3   f-k Filter -- Relative Units

We said earlier that the f-k filter slopes can be specified in relative units of time over trace spacing (for example, seconds/meter) or in absolute units of time samples per trace. In this section, we specify the f-k filter slopes in relative units: we measure the vertical axis as time in seconds and we measure the horizontal axis as distance in meters.

Again, we use time- and trace-windowed Oz gather 8. Below is the **surange** output of *oz08w.su*.

```
surange  <  oz08w.su
```

```
71 traces:
 tracl=(26,96)  tracr=(26,96)  fldr=10008 tracf=(26,96)  cdp=(33,103)
 cdpt=1 trid=1 nvs=1 nhs=1 duse=1
 scalel=1 scalco=1 counit=1 delrt=4 muts=4
 ns=750 dt=4000
```

From **surange**, we know the sample interval is 0.004 seconds (dt=4000). From Table 1-8 of Seismic Data Processing (Yilmaz, 1987) or Table 1-13 of Seismic Data Analysis (Yilmaz, 2001), we know the trace spacing is 50 meters.

Table 14.1: Absolute and relative filter slope values

Absolute	samples/trace	4	5	6	7
Relative	**seconds/meter**	0.016/50	0.020/50	0.024/50	0.028/50
		0.00032	0.00040	0.00048	0.00056

In script *ifk.sh*, we change lines 21-22:

```
20  # Processing variables [ Instructions below ]
21  dx=50  # trace spacing
22  dt=0.004  # time sample interval
```

Below is the screen dialog with our input highlighted. Line numbers are added for discussion. We input values on lines 14, 21, and 27. We supply our slope values (relative units) on dialog line 21.

```
 1  ----------------------------------------------------
 2                   f-k Filter Test
 3                   --------------
 4   From your slope values, two filters are created --
 5        a pass filter and a reject filter.
 6   When you exit, the following files are output:
 7             .  Slopes are in  ==>  fk.txt
 8            Passed data are in  ==>  fkpass.su
 9          Rejected data are in  ==>  fkrejj.su
10  ----------------------------------------------------
11
12  Supply gain power value for t^(power)
13  For no gain, supply 0
14  2
15
16  Supply filter slopes.
17  Input: s1,s2,s3,s4    where s1 < s2 < s3 < s4
18     Example:  3.0,3.5,4.0,4.5
19         or:  -4.5,-4.0,-3.5,-3.0
20  Use commas. Do not use spaces.
21  .00032,.00040,.00048,.00056
22
23            Slopes: .00032,.00040,.00048,.00056
24
25  Enter 1 for more f-k filter testing
26  Enter 2 to EXIT
27  2
28
29  Processing log is in  ==>  fk.txt
30    Passed data are in  ==>  fkpass.su
31  Rejected data are in  ==>  fkrejj.su
32
```

```
33  press return key to exit
```

Figure 14.3 shows the input time gather and its f-k equivalent.

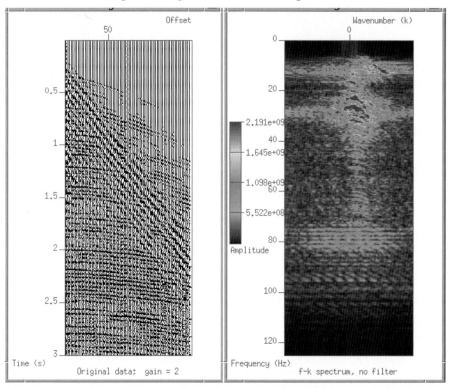

Figure 14.3: *Input gather oz08w.su. Left: Seismic after gain (tpow=2.0). Right: f-k plot of seismic after gain.*

Figure 14.4 shows the four output plots: a t-x plot and an f-k plot after application of the pass filter and a t-x plot and an f-k plot after application of the reject filter. (The test can be repeated).

We rename the output files to prevent our files from becoming overwritten by a later run: **Note:** Use a semicolon to separate multiple commands on the same line.

mv fkpass.su fkpass08rel.su; mv fkrejj.su fkrejj08rel.su

We are happy to see that Figure 14.2 and Figure 14.4 look the same. We compare these data sets in the next section.

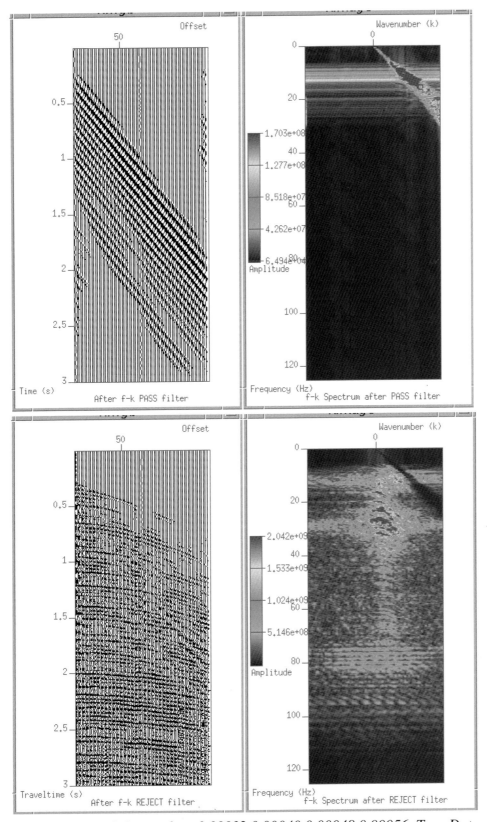

Figure 14.4: *Output of slope values 0.00032, 0.00040, 0.00048, 0.00056. Top: Data passed by filter. Bottom: Data rejected by filter.*

14.4 f-k Filter -- Discussion

First, we discuss the amplitudes of the two methods used above. Second, we discuss the vertical and horizontal annotations on the f-k plots.

Starting with the reject files from above, we made a difference file:

```
sudiff fkrejj08abs.su fkrejj08rel.su > fkrejj08diff.su
```

Then, from all three data sets, we windowed times 1.000 to 1.040 from trace 50. The amplitudes of the two methods are very similar, and the difference amplitudes are much smaller than those of the absolute and relative files.

Table 14.2: Reject files, trace 50

Time	Absolute	Relative	Difference
1.000	2.1742e+06	2.1742e+06	4.0000e+00
1.004	1.9120e+06	1.9120e+06	2.8750e+00
1.008	1.1568e+06	1.1568e+06	1.0000e+00
1.012	-8.8561e+05	-8.8561e+05	-5.0000e-01
1.016	-2.5479e+06	-2.5479e+06	-2.0000e+00
1.020	-2.0101e+06	-2.0101e+06	-3.1250e+00
1.024	-4.7880e+05	-4.7880e+05	-4.2812e+00
1.028	1.9655e+03	1.9688e+03	-3.2500e+00
1.032	-3.0930e+05	-3.0930e+05	-4.2500e+00
1.036	-4.5027e+05	-4.5026e+05	-3.3125e+00
1.040	-1.5578e+05	-1.5577e+05	-3.3125e+00

If we plot only the difference file, it is difficult to see that its amplitudes are much smaller than those of the absolute and relative files.

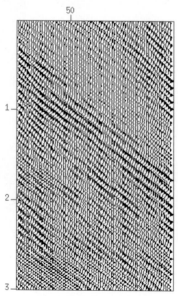

Figure 14.5: *File fkrejj08diff.su.*

Using the following commands, we concatenate the three files with the difference file in the middle. The first command creates the file that eventually holds all three files.

```
cat  fkrejj08abs.su  >   fkrejj08BIG.su
cat  fkrejj08diff.su  >>  fkrejj08BIG.su
cat  fkrejj08rel.su  >   fkrejj08BIG.su
```

Then we plot the result.

```
suxwigb  <  fkrejj08BIG.su  perc=99  &
```

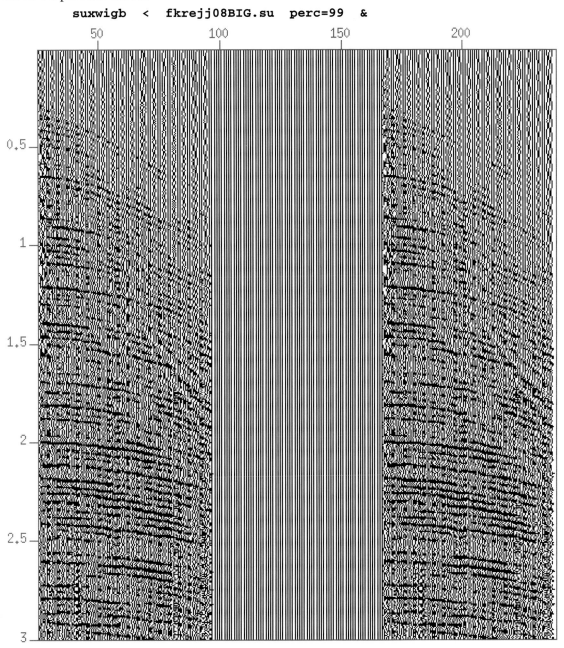

Figure 14.6: *File fkrejj08BIG.su, the concatenation of the absolute method reject file (left),the relative method reject file (right), and the difference of the two (middle).*

In Figure 14.6, we see that the amplitudes of the difference file are much smaller than the amplitudes of the other two files. Figure 14.6 shows vertical bands because the computer screen cannot clearly show many traces squeezed into a narrow space.

To understand the vertical and horizontal annotations on the f-k plots, remember the definition of the Nyquist frequency. The Nyquist frequency, introduced in Section 10.4, is the folding frequency. Mathematically, it is the inverse of twice the sample interval. Units of temporal frequency (f) are the number of cycles per second. Units of spatial frequency (wavenumber or k) are the number of cycles per unit distance. In Table 14.3, we present the distance and time Nyquist frequencies for the absolute and relative examples we used earlier.

Table 14.3: Nyquist spatial and temporal frequencies.

	Absolute		**Relative**	
ifk.sh	dx=1	dt=1	dx=50	dt=0.004
Nyquist	$k_N = 0.5$	$f_N = 0.5$ Hz	$k_N = 0.01$	$f_N = 125$ Hz

Figure 14.7 is a stretched version of Figure 14.1, right. Here, the horizontal annotations are easier to see.

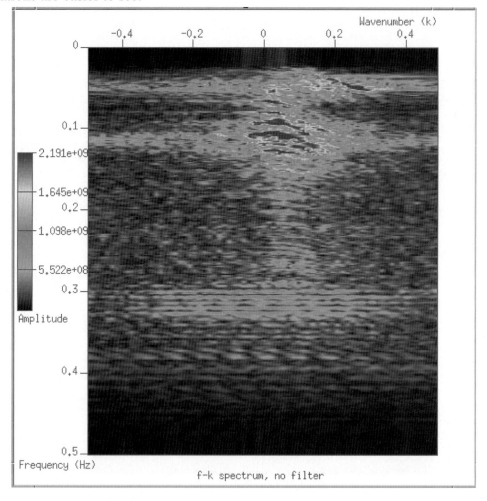

Figure 14.7: *An f-k plot of the input data, made using absolute dx and dt values.*

Figure 14.8 is a stretched version of Figure 14.3, right. Here again, the horizontal annotations are easier to see.

For both Figures 14.7 and 14.8, the maxima of the axes (negative and positive) are the respective Nyquist frequencies, spatial and temporal.

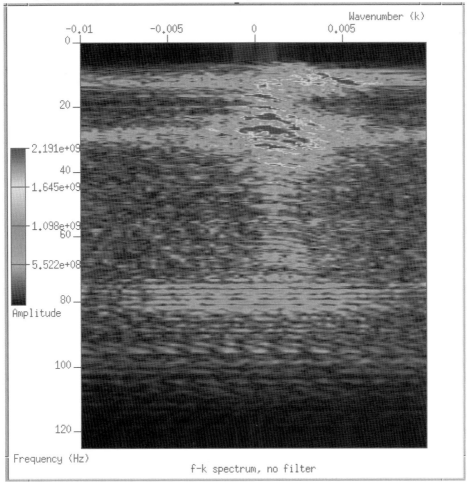

Figure 14.8: *An f-k plot of the input data, made using relative dx and dt values.*

14.5 Deconvolution

"Deconvolution is a process that improves the temporal resolution of seismic data by compressing the basic seismic wavelet (Yilmaz, 1987, Yilmaz 2001: Chapter 2)." The two important parameters of deconvolution in **supef** (Wiener predictive error filtering) are *minlag* (seconds) and *maxlag* (seconds). Our deconvolution script, *idecon*, refers to these as prediction length and operator length (respectively). We take this terminology from Robinson and Treitel (2000). In explaining Figure 14.9, Robinson and Treitel (2000) write, "Empirical studies indicate that α [prediction length] should be set roughly equal to the lag that corresponds to the second zero crossing of the autocorrelation function."

Note: Operator length must be longer than prediction length.

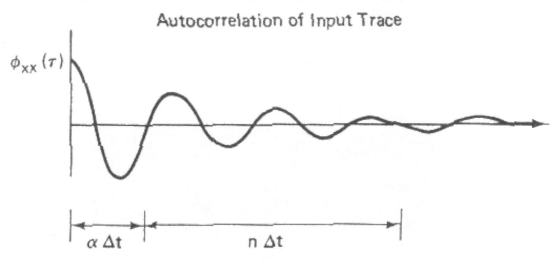

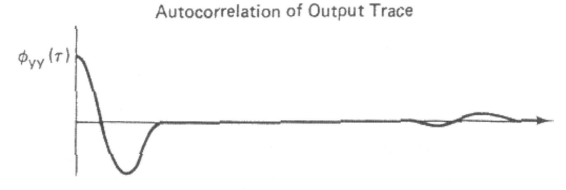

Figure 14.9: *Top: Autocorrelation of trace with reverberation. Bottom: Same trace after deconvolution using prediction distance α and operator length n (Robinson and Treitel, 2000, Figure 12-3). Used with permission.*

14.5.1 Use Deconvolution Script idecon

For our interactive deconvolution processing example, we use Oz gather 16 from the CSM CWP web site (Section 3.3). Below is the **surange** output of *oz16.su*.

```
        surange < oz16.su
48 traces:
 tracl=(1,48)  tracr=(1,48)  fldr=10016 tracf=(1,48)  cdp=(16,63)
 cdpt=1 trid=1 nvs=1 nhs=1 duse=1
 scalel=1 scalco=1 counit=1 delrt=4 muts=4
 ns=1325 dt=4000
```

We want to use only 3 seconds of this 4 millisecond sample interval data set, and we want to mute the strong refractions. We examine the data using the following line command:

```
        sugain < oz16.su tpow=2 | suxwigb perc=95 &
```

The following line command mutes the early arrivals and windows the first 3 seconds:

```
sumute < oz16.su key=tracl tmute=1.200,0.395 xmute=1,48 | suwind tmax=3 >
oz16m3.su
```

Below is the **surange** output of *oz16m3.su*.

```
        surange < oz16m3.su
48 traces:
 tracl=(1,48)  tracr=(1,48)  fldr=10016 tracf=(1,48)  cdp=(16,63)
 cdpt=1 trid=1 nvs=1 nhs=1 duse=1
 scalel=1 scalco=1 counit=1 delrt=4 muts=(395,1200)
 ns=750 dt=4000
```

As we stated in the introduction, script *idecon* uses programs **supef** and **suacor**. Program **supef** does the deconvolution. Script idecon uses **suacor** to make plots of each trace's autocorrelation, an important deconvolution analysis tool. Below is the parameter list from the **supef** selfdoc:

```
SUPEF - Wiener predictive error filtering

 supef <stdin >stdout   [optional parameters]

 Required parameters:
        dt is mandatory if not set in header

 Optional parameters:
        minlag=dt               first lag of prediction filter (sec)
        maxlag=last             lag default is (tmax-tmin)/20
        pnoise=0.001            relative additive noise level
        mincorr=tmin            start of autocorrelation window (sec)
        maxcorr=tmax            end of autocorrelation window (sec)
        showwiener=0            =1 to show Wiener filter on each trace
        mix=1,...               array of weights (floats) for moving
                                average of the autocorrelations
```

Notice that **supef** has windowing parameters for the autocorrelation (parameters *mincorr* and *maxcorr*).

Below is the **suacor** selfdoc:

```
SUACOR - auto-correlation

 suacor <stdin >stdout [optional parms]

 Optional Parameters:
 ntout=101       odd number of time samples output
 norm=1 if non-zero, normalize maximum absolute output to 1
 sym=1           if non-zero, produce a symmetric output from
                        lag -(ntout-1)/2 to lag +(ntout-1)/2
```

Notice that **suacor** does not have any windowing parameters.

In *idecon*, if the user decides to window the autocorrelation prior to applying deconvolution, *idecon* windows the traces that go into **suacor** (that go into the autocorrelation plot).

There are only two values that must be supplied before using *idecon*: the name of the seismic file and the percent white noise for **supef** (**supef** parameter *pnoise*). All other parameters are input in response to questions that appear on the screen. Below is the screen dialog with our input highlighted. Line numbers are added for discussion. We input values on lines 8, 14, 18, 22, 37, 40, 42, 50, 52, 56, 60, 74, 77, 79, 87, 89, and 93.

```
 1
 2      *** DECONVOLUTION TEST ***
 3
 4      ENVIRONMENT QUESTIONS ...
 5
 6    Supply gain power value for t^(power)
 7    For no gain, supply 0
 8    2
 9
10    Supply "perc" value for plots
11       Typical values are 100, 98, 95, 90, 85
12       If you need a perc value less than 70,
13       your data are probably poorly gained.
14    98
15
16    Supply "key" for plot x-axis annotation
17       Typical values are:  tracl  tracf  offset
18    tracl
19
20    For wiggle traces, enter w
21    For image display, enter i
22    w
23
24      ITERATION QUESTIONS ...
25
26        oz16m3.su:
27
28      ---------------------------------------
29        Sample interval is .004000 seconds
30        Data last time is 3.000000 seconds
31      ---------------------------------------
32
33        Autocorrelation window start time is .004000
34        Autocorrelation window last time is 3.000000
35
36    Change autocorrelation analysis window ? (y/n)
37    y
38
39    Supply first time (seconds)
40    1.
41    Supply last time (seconds)
42    2.8
43
44        oz16m3.su:
45
46        Autocorrelation first time is 1.
47        Autocorrelation last time is 2.8
48
49    Enter prediction length (seconds)
50    .030
51    Enter operator length (seconds)
52    0.175
```

```
53
54   Enter 1 for more decon testing
55   Enter 2 for EXIT
56   1
57
58   Enter R to re-process deconvolved data
59   Enter S to start over
60   r
61    -> Using modified data
62
63      oz16m3.su:
64
65      ---------------------------------------
66      Sample interval is .004000 seconds
67       Data last time is 3.000000 seconds
68      ---------------------------------------
69
70      Autocorrelation window start time is 1.
71      Autocorrelation window last time is 2.8
72
73   Change autocorrelation analysis window ? (y/n)
74   y
75
76   Supply first time (seconds)
77   .004
78   Supply last time (seconds)
79   3.000
80
81      oz16m3.su:
82
83      Autocorrelation first time is .004
84      Autocorrelation last time is 3.000
85
86   Enter prediction length (seconds)
87   .004
88   Enter operator length (seconds)
89   .040
90
91   Enter 1 for more decon testing
92   Enter 2 for EXIT
93   2
94
95   Output seismic file is  ==>  idecon.su
96      Output log file is  ==>  idecon.txt
97
98   press return key to exit
99
```

In the beginning, *idecon* asks for "environment" values: a gain value, a *perc* plot value, a key for labeling the **suxwigb** x-axis, and whether the user wants to see wiggle plots or image plots. The next questions are asked with each loop through *idecon*.

Dialog lines 29 and 30 are printed with each loop to remind the user of the sample interval and the total time of the data. By knowing the sample interval, the user knows which value to supply for spiking deconvolution. Initially, these two values are used to set the autocorrelation window (dialog lines 33-34). Before the first autocorrelation plot is made, the user is given the chance to change start and stop time of the autocorrelation window (line 36). In this case, the user changes the values (dialog lines 39-42). The user is immediately given a confirmation message of the autocorrelation window values. At this time, the right side of Figure 14.10 is made.

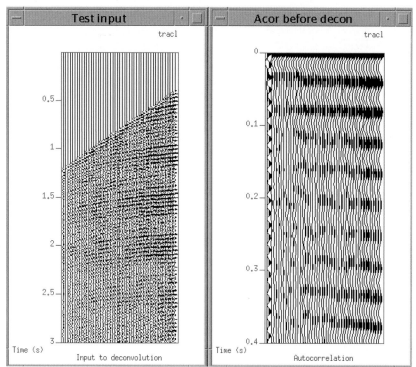

Figure 14.10: *Left: Seismic data for deconvolution. Right: Autocorrelation of the seismic traces. In this first decon test, the autocorrelation plot was made using seismic traces windowed (**suwind**) from 1 second to 2.8 seconds. (See dialog lines 36-42).*

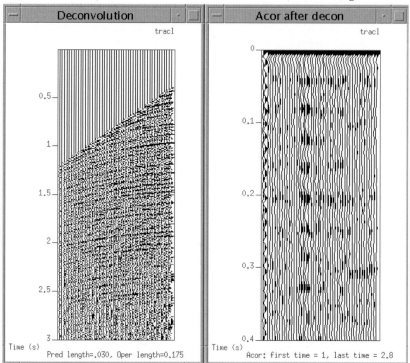

Figure 14.11: *Left: Seismic data after deconvolution. Right: Autocorrelation of the windowed portion of the deconvolved seismic traces.*

After the user supplies a prediction length and an operator length (dialog lines 49-52), the plots of Figure 14.11 are made (a gapped deconvolution test).

For the second test, the user decides to deconvolve the output of the first test (instead of using the original input (dialog lines 58-61). The user is reminded of the seismic file's sample interval and total time (dialog lines 66-67) as well as the current settings for the autocorrelation window (dialog lines 70-71).

Again, the user decides to change the autocorrelation window, supplying values that correspond to the entire trace (dialog lines 73-79).

For the new test, the user supplies a prediction length of one sample and an operator length (dialog lines 86-89), and the plots of Figure 14.12 are made for a "spiking" deconvolution. The autocorrelation plot of Figure 14.12 is made from the entire trace.

The seismic plots show the prediction length and the operator length, while the autocorrelation plots show the extent of the autocorrelation windows.

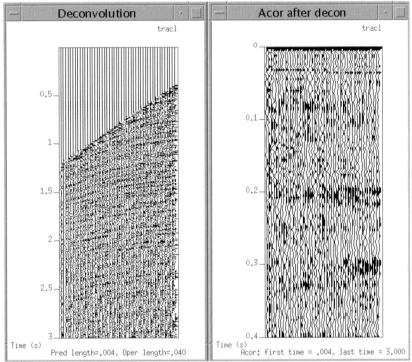

Figure 14.12: *Left: Seismic data after spiking deconvolution. Right: Autocorrelation of the deconvolved seismic traces (using the entire times).*

Below are the contents of *idecon.txt*, the log file. Line numbers are added for discussion. Line 2 includes the percent white noise and the gain (t^(power)) values.

```
 1    *** DECONVOLUTION TEST ***
 2       Input file = oz16m3.su
 3       Percent white noise = 0.001    Gain = 2
 4       Sample interval is .004000 seconds
 5        Data last time is 3.000000 seconds
 6       -----------------------
 7    -> Original data
 8       Autocorrelation first time is 1.
 9       Autocorrelation last time is 2.8
10       Prediction length = .030
11       Operator length = 0.175
12    -> Using modified data
13       Autocorrelation first time is .004
14       Autocorrelation last time is 3.000
15       Prediction length = .004
```

```
16      Operator length = .040
```

14.5.2 Script idecon.scr

We usually start interactive scripts by running a *.scr* script. Here is *idecon.scr*:

```
1  #! /bin/sh
2  # File: idecon.scr
3  #       Run this script to start script idecon.sh
4
5  xterm -geom 60x18+10+544 -e idecon.sh
6
```

Line 1 invokes the shell. Line 5 opens a dialog window 80 characters wide by 18 characters high. Line 5 also places the dialog window 10 pixels from the left side of the viewing area and 544 pixels from the top of the viewing area. Line 5 also starts *idecon.sh*, the processing script, within that window.

14.5.3 Script idecon.sh

We start interactive scripts by running a *.scr* script. Here is *idecon.sh*.

```
1   #! /bin/sh
2   # File: idecon.sh
3   #       Run script idecon.scr to start this script
4   #       You can window the seismic data for autocorrelation
5   #         analysis. For predictive deconvolution, based on
6   #         the autocorrelation display, use a prediction length
7   #         equal to the time from zero to the second zero crossing.
8   #       Example:
9   #               Prediction length = .030 seconds
10  #               Operator length = .250 seconds
11
12  # Set messages on
13  ##set -x
14
15  #================================================
16  # User-supplied values
17  #------------------------------------------------
18
19  # Input seismic file
20  indata=oz16m3.su
21
22  # percent white noise [supef default=0.001]
23  wnoise=0.001
24
25  #================================================
26
27  # DOCUMENT TERMS:
28  #
29  #    PL = Prediction length
30  #    OL = Operator length
31  #
32  #  |
33  #  **
34  #  | *                **
35  #  |  *             *   *          **
36  #  |   *          *      *        *   *     **
37  #  ------*------*----  *-----*--  *--*--*--*
38  #  |      *      *|       *   *       **    |**
39  #  |       *   *  |        **              |
40  #  |        **    |                        |
41  #  |            |                        |
```

```
42  #  |  == PL ===  |  ======= OL =======  |
43  #
44  #   Input wavelet
45  #
46  #
47  #  |
48  #  **
49  #  |  *
50  #  |   *
51  #  |    *
52  #  ------*------*********************--*
53  #  |      *     *                    **
54  #  |        *   *
55  #  |         **
56  #  |
57  #
58  #   Output wavelet
59  #
60  #
61  # REFERENCE: Robinson, E.A., Treitel, S., 2000, Geophysical
62  #            Signal Analysis, Society of Exploration
63  #            Geophysicists, Tulsa, p. 278-279.
64  #
65  #-------------------------------------------------
66  # Describe temporary files
67  #-------------------------------------------------
68
69  # tmp0 = input seismic file, after gain
70  # tmp1 = input to deconvolution:
71  #          tmp0 (reproc=s) or tmp3 (reproc=r)
72  # tmp2 = input seismic windowed for autocorrelation
73  # tmp3 = output of the prediction error filter
74  # tmp4 = output of the prediction error filter
75  #          windowed for autocorrelation
76  # tmp5 = ASCII log file of processing parameters
77  # tmp6 = ASCII file to reduce screen display of "zap"
78
79  #-------------------------------------------------
80
81  echo " "
82  echo "  *** DECONVOLUTION TEST ***"
83  echo " "
84  echo "  ENVIRONMENT QUESTIONS ..."
85
86  # Remove temporary files
87  rm -f tmp*
88
89  #-------------------------------------------------
90  # Supply gain value to t^(power)
91  # Supply perc value for plots
92  # Supply key value for plots
93
94  echo " "
95  echo "Supply gain power value for t^(power)"
96  echo "For no gain, supply 0"
97  > /dev/tty
98  read tpow
99
100 echo " "
101 echo "Supply \"perc\" value for plots"
102 echo "  Typical values are 100, 98, 95, 90, 85"
103 echo "  If you need a perc value less than 70,"
104 echo "  your data are probably poorly gained."
```

```
105   > /dev/tty
106   read myperc
107
108   echo " "
109   echo "Supply \"key\" for plot x-axis annotation"
110   echo " Typical values are:  tracl  tracf  offset"
111   > /dev/tty
112   read mykey
113
114   #---------------------------------------------------
115   # Ask: wiggle traces (xwigb) or image (ximage)
116
117   echo " "
118   echo "For wiggle traces, enter w"
119   echo "For image display, enter i"
120   > /dev/tty
121   read display
122
123   case $display in
124     i)
125        wiggle=1
126        ;;
127     *)
128        wiggle=0
129        ;;
130   esac
131
132   #---------------------------------------------------
133   # Apply gain, copy input data to temporary file
134
135   sugain < $indata tpow=$tpow > tmp0
136
137   #---------------------------------------------------
138   # From data: get number of samples, sample interval
139   # Compute time length of traces
140
141   ns=`sugethw ns < tmp0 | sed 1q | sed 's/.*ns=//'`
142   dt=`sugethw dt < tmp0 | sed 1q | sed 's/.*dt=//'`
143
144   # Convert dt key value from microsec to sec
145   tsamp=`bc -l << -END
146     scale=6
147     $dt / 1000000
148   END`
149
150   # Compute last time (seconds)
151   tend=`bc -l << -END
152     $ns * $tsamp
153   END`
154
155   #---------------------------------------------------
156   # Write to log file
157
158   echo "  *** DECONVOLUTION TEST ***" > tmp5
159   echo "    Input file = $indata" >> tmp5
160   echo "    Percent white noise = $wnoise   Gain = $tpow" >> tmp5
161   echo "    Sample interval is $tsamp seconds" >> tmp5
162   echo "     Data last time is $tend seconds" >> tmp5
163   echo "   -------------------------" >> tmp5
164
165   #---------------------------------------------------
166
167   # initial setting for autocorrelation analysis window start
```

```
294      suximage perc=$myperc xbox=322 ybox=10 wbox=300 hbox=500 \
295              label1=" Time (s)" \
296              windowtitle="Acor before decon" \
297              title="Autocorrelation" wclip=0 verbose=0 &
298    fi
299
300  #------------------------------------------------
301  # Prediction length & Operator length
302
303    echo " "
304    echo "Enter prediction length (seconds)"
305    > /dev/tty
306    read minlag
307    echo "Enter operator length (seconds)"
308    > /dev/tty
309    read maxlag
310
311    echo "   Prediction length = $minlag" >> tmp5
312    echo "   Operator length = $maxlag" >> tmp5
313
314  #------------------------------------------------
315  # Deconvolution
316
317    supef < tmp1 minlag=$minlag maxlag=$maxlag \
318           mincorr=$taa maxcorr=$tzz pnoise=$wnoise > tmp3
319
320  #------------------------------------------------
321  # After deconvolution -- plot data
322
323    if [ $wiggle -eq 0 ] ; then
324      suxwigb < tmp3 perc=$myperc xbox=634 ybox=10 wbox=300 hbox=500 \
325              label1=" Time (s)" label2=$mykey key=$mykey \
326              title="Pred length=$minlag, Oper length=$maxlag" \
327              windowtitle="Deconvolution" verbose=0 &
328    else
329      suximage < tmp3 perc=$myperc xbox=634 ybox=10 wbox=300 hbox=500 \
330              label1=" Time (s)" \
331              title="Pred length=$minlag, Oper length=$maxlag" \
332              windowtitle="Deconvolution" verbose=0 &
333    fi
334
335  #------------------------------------------------
336  # After deconvolution -- plot autocorrelation
337
338    suwind < tmp3 tmin=$taa tmax=$tzz > tmp4
339
340    if [ $wiggle -eq 0 ] ; then
341      suacor <tmp4 ntout=101 sym=0 |
342      suxwigb perc=$myperc xbox=946 ybox=10 wbox=300 hbox=500 \
343              label1=" Time (s)" label2=$mykey key=$mykey \
344              title="Acor: first time = $taa, last time = $tzz" \
345              windowtitle="Acor after decon" verbose=0 &
346    else
347      suacor < tmp4 ntout=101 sym=0 |
348      suximage perc=$myperc xbox=946 ybox=10 wbox=300 hbox=500 \
349              label1=" Time (s)" \
350              title="Acor: first time = $taa, last time = $tzz" \
351              windowtitle="Acor after decon" verbose=0 &
352    fi
353
354  #------------------------------------------------
355  # Do more decon or exit
356
```

```
357    echo " "
358    echo "Enter 1 for more decon testing"
359    echo "Enter 2 for EXIT"
360    > /dev/tty
361    read selection
362
363      case $selection in
364         1)
365             finish=false
366             ;;
367         2)
368             cp tmp3 idecon.su
369             cp tmp5 idecon.txt
370             echo " "
371             echo "Output seismic file is  ==>  idecon.su"
372             echo "   Output log file is  ==>  idecon.txt"
373             pause exit
374             zap xwigb > tmp6  # decrease screen messages
375             zap ximage > tmp6  # decrease screen messages
376             finish=true
377             ;;
378      esac
379
380    new=false
381
382  done
383
384  #-------------------------------------------------
385  # Exit
386  #-------------------------------------------------
387
388  # Remove temporary files
389  rm -f tmp*
390
391  # Exit politely from shell
392  exit
393
```

Script idecon.sh has the following major sections:

- Lines 19-23 are user-supplied values: the input file and a percent white noise value.

- Lines 27-64 describe prediction length and operator length and provide a reference. Also, lines 66-77 describe the temporary files.

- Lines 81-182 operate before the deconvolution test loop.

  o Line 87 removes old temporary files.

  o Lines 95-98 get the user's gain value.

  o Lines 101-106 get the user's *perc* value.

  o Lines 109-112 get the users' key plot value.

  o Lines 118-130 get the user's preference for wiggle or image plots.

  o Line 135 applies gain to the input data and holds the output in *tmp0*.

  o Lines 141-153 compute the sample interval (seconds) and the time length of the traces (seconds).

  o Lines 158-163 write information to the log file.

- o Lines 167-170 create initial autocorrelation window analysis start and stop times – the values computed in lines 141-153.
- o Line 174 sets a flag that means this is the first time through the test loop.
- o Line 175 sets a flag that means the user is not finished testing.
- Lines 185-382 are the test loop.
  - o Lines 187-190: If this is the first time the user is in the test loop (new=true), the contents of *tmp0* are copied to *tmp1*. File *tmp1* is the file that is displayed (lines 220-230) and windowed for autocorrelation analysis and display (lines 284-298). The windowed contents of *tmp1* are output to *tmp2*. File *tmp2* is converted to autocorrelation traces and displayed (lines 286-291 or lines 293-297).
  - o Lines 192-211: If this is not the first test (new=false), the user chooses whether to re-process the deconvolved data (*tmp3*) or process the original data (*tmp0*).
  - o Line 215 removes the two files that are created after deconvolution. File *tmp3* holds the deconvolution output (made by **supef**, lines 317-318). Files *tmp4* holds the deconvolved and time-windowed data (line 338) that are converted to autocorrelation traces and displayed (lines 340-345 or lines 346-351).
  - o Lines 233-279 inform the user of the current autocorrelation window limits (lines 236-244), allow the user to set new limits (lines 250-271) , and again inform the user of the autocorrelation window limits (lines 276-279).
  - o After the test seismic data are displayed and the autocorrelation traces are displayed, the user is asked for prediction length and operator length (lines 304-309). The user-supplied values are written to the log file (lines 311-312).
  - o After the deconvolution results and the post-deconvolution autocorrelation traces are displayed, the user chooses whether to run a new test or exit the script (lines 358-361). If the user chooses to exit (lines 367-377), the last seismic deconvolution output data are written to a disk file (line 368), the temporary log file is written to a disk file (line 369), screen messages inform the user of the names of the disk files (lines 371-372), the displays are closed (lines 374-375), and the looping flag is set to end the loop (finish=true).
  - o Line 380 sets a flag, new=false. This means the user will always have to choose whether the next test re-processes the deconvolved data or uses the original data (lines 192-211). Line 380 is at the bottom of the loop, but could be placed just after the new test, on line 214.

We encourage you to try these tools on the Nankai and Taiwan data sets.

15. Taiwan: Sort, Velocity Analysis, NMO, Stack

15.1 Introduction

Returning to the Taiwan data, we now have 5 seconds of data, we zeroed the bad hydrophones, applied spherical divergence correction, and band-pass filtered the shot gathers. The result is *Tshot5kgf.su*. We will now

- sort these shot gathers to CMP order,
- analyze the CMPs for stacking velocities,
- use the stacking velocities to apply NMO, and
- stack the CMPs.

15.2 Sort

Sorting to CMP order is a one-line command:

```
susort  <  Tdata/Tshot5kgf.su  cdp  offset  >  Tdata/Tcmp5kgf.su
```

We send the sorted data to "Tdata" since this file is just as large as the file of windowed (0-5 seconds) shot gathers: 133 Mbytes.

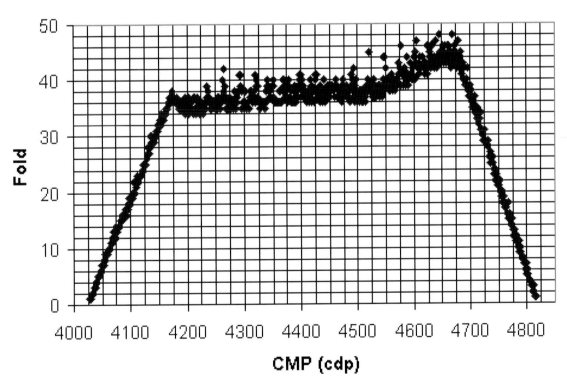

Figure 15.1: *Fold values for data set Tcmp5kgf.su.*

Below is the **surange** output of *Tcmp5kgf.su*.

```
surange  <  Tcmp5kgf.su
25344 traces:
```

```
tracl=(1,25344)   tracr=(1,25344)   fldr=(800,975)   tracf=(1,144)   ep=(740,915)
cdp=(4027,4816)   cdpt=(1,144)   trid=1 nhs=1 offset=(-3663,-88)
sdepth=80000 swdep=(2120000,3540000)   scalel=-10000 scalco=1 sx=(-4773100,-
3965250)
gx=(-4764600,-3600343)   gy=(766,94805)   counit=3 tstat=12 ns=1251
dt=4000 gain=9 afilf=160 afils=72 lcf=3
hcf=160 lcs=6 hcs=72 year=95 day=260
hour=(12,13)   minute=(0,59)   sec=(0,59)
```

We have 790 CMP gathers (4816-4027+1). Figure 15.1 shows that the full fold gathers (at least 34 traces per gather) are approximately 4175-4700. (We describe the program, **sukeycount,** which generated the data for Figure 15.1 in Appendix C.) Despite the constant number of traces per shot, CMP fold varies (Figure 15.1) because ship speed and streamer position varied.

15.3   Velocity Analysis

Below is the user area of *iva.sh* changed for the Taiwan data. To make reasonably detailed velocity analysis at a regular interval along the line, we choose to analyze every 25th CMP starting with CMP 4175 (lines 13-18). We set the *perc* value to 90 (line 31), an accommodation for real data. Semblance values go from 1000 m/s (line 38) to 5000 m/s (nvs*dvs + fvs). CVS panels range from 1000 m/s to 5000 m/s (lines 43-46).

```
 8    #===================================================
 9    # USER AREA -- SUPPLY VALUES
10    #---------------------------------------------------
11    # CMPs for analysis
12
13     cmp1=4175   cmp2=4200   cmp3=4225   cmp4=4250
14     cmp5=4275   cmp6=4300   cmp7=4325   cmp8=4350
15     cmp9=4375 cmp10=4400 cmp11=4425 cmp12=4450
16    cmp13=4475 cmp14=4500 cmp15=4525 cmp16=4550
17    cmp17=4575 cmp18=4600 cmp19=4625 cmp20=4650
18    cmp21=4675 cmp22=4700
19
20    numCMPs=22
21
22    #---------------------------------------------------
23    # File names
24
25    indata=Tdata/Tcmp5kgf.su  # SU format
26    outpicks=Tvpick.txt  # ASCII file
27
28    #---------------------------------------------------
29    # display choices
30
31    myperc=90        # perc value for plot
32    plottype=1       # 0 = wiggle plot,  1 = image plot
33
34    #---------------------------------------------------
35    # Processing variables
36
37    # Semblance variables
38    nvs=160  # number of velocities
39    dvs=25   # velocity intervals
40    fvs=1000 # first velocity
41
42    # CVS variables
43    fc=1000 # first CVS velocity
44    lc=5000 # last CVS velocity
45    nc=10   # number of CVS velocities (panels)
```

```
46  XX=11    # ODD number of CMPs to stack into central CVS
47
48  #================================================
```

We make one more accommodation for real data. When we use **suximage** following **suvelan**, we set *perc=97* (highlighted below). These two lines occur twice, once for plotting the velan with picks overlain and once for plotting the original velan. See lines 305 and 316 in Section 7.6.6.C.

```
suvelan < panel.$picknow nv=$nvs dv=$dvs fv=$fvs |
suximage xbox=10 ybox=10 wbox=300 hbox=450 perc=97 \
```

By changing *perc* from 99 (in Section 7.6.6.C) to 97, we "increase the volume" of the semblance values.

Below are two of the semblance plots with picks overlain. Our philosophy was to pick generally increasing velocities. Better picks might be made by following subtle events. We encourage you to experiment.

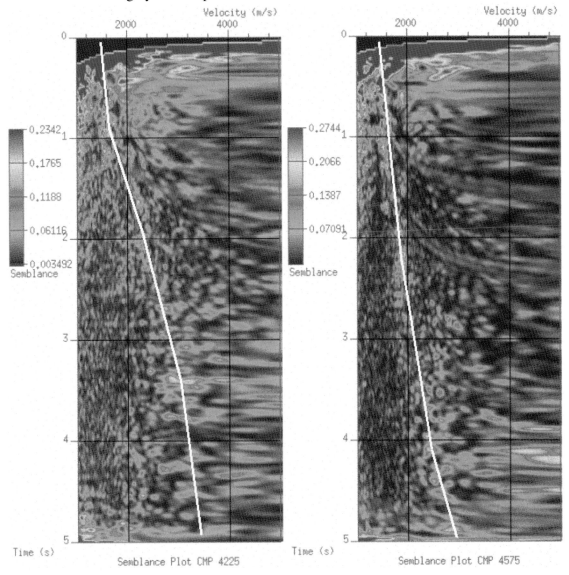

Figure 15.2: *Semblance plots. Left: CMP 4225. Right: CMP 4575.*

Below is the file of t-v picks. For the sake of presentation, we made line 1 into two lines. Line 1 as presented here has no spaces at the end of the first part and no spaces at the beginning of the second part, making this a usable file. Remember to remove line 2 before using this file in **sunmo**.

```
 1  cdp=4175,4200,4225,4250,4275,4300,4325,4350,4375,4400,4425,\
    4450,4475,4500,4525,4550,4575,4600,4625,4650,4675,4700 \
 2  #=1,2,3,4,5,6,7,8,9,10,11,12,13,14,15,16,17,18,19,20,21,22 \
 3  tnmo=0.0527704,0.527704,2.26913,3.219,3.7467,4.96042 \
 4  vnmo=1485.83,1574.17,2325,2634.17,3252.5,4224.17 \
 5  tnmo=0.0659631,1.0686,3.33773,5 \
 6  vnmo=1463.75,1706.67,3053.75,3826.67 \
 7  tnmo=0.0527704,0.844327,2.00528,3.37731,4.93404 \
 8  vnmo=1507.92,1640.42,2325,3031.67,3426.41 \
 9  tnmo=0.0395778,0.897098,2.41425,3.49604,4.97361 \
10  vnmo=1530,1640.42,2347.08,3009.58,4069.58 \
11  tnmo=0.0527704,0.620053,2.71768,3.64116,4.98681 \
12  vnmo=1463.75,1596.25,2479.58,2832.92,3583.75 \
13  tnmo=0.0395778,0.633245,2.62533,3.48285,4.98681 \
14  vnmo=1463.75,1596.25,2656.25,3208.33,3892.92 \
15  tnmo=0.0527704,0.633245,3.7467,4.98681 \
16  vnmo=1441.67,1552.08,3451.25,3848.75 \
17  tnmo=0.0527704,0.699208,2.8496,4.98681 \
18  vnmo=1507.92,1662.5,2567.92,3053.75 \
19  tnmo=0.0263852,1.04222,3.85224,4.97361 \
20  vnmo=1441.67,1772.92,3164.17,3716.25 \
21  tnmo=0.0527704,1.22691,3.10026,4.12929,5 \
22  vnmo=1463.75,1728.75,2590,3539.58,3804.58 \
23  tnmo=0.0527704,1.18734,2.4934,3.66755,5 \
24  vnmo=1463.75,1684.58,2413.33,2899.17,3252.5 \
25  tnmo=0.0527704,1.10818,2.94195,3.69393,4.97361 \
26  vnmo=1485.83,1706.67,2413.33,2921.25,3296.67 \
27  tnmo=0.0527704,0.91029,3.68074,4.98681 \
28  vnmo=1507.92,1772.92,3495.42,4091.67 \
29  tnmo=0.0659631,1.21372,3.69393,4.97361 \
30  vnmo=1530,1596.25,2965.42,3738.33 \
31  tnmo=0.0527704,1.21372,3.44327,5.01319 \
32  vnmo=1485.83,1662.5,2766.67,3694.17 \
33  tnmo=0.0395778,1.26649,3.79947,4.97361 \
34  vnmo=1463.75,1640.42,2877.08,3097.92 \
35  tnmo=0.0527704,2.24274,4.1161,4.97361 \
36  vnmo=1463.75,1883.33,2457.5,2943.33 \
37  tnmo=0.0395778,1.93931,3.25858,5.01319 \
38  vnmo=1441.67,1927.5,2457.5,3451.25 \
39  tnmo=0.0659631,1.27968,3.25858,4.97361 \
40  vnmo=1485.83,1618.33,2810.83,3826.67 \
41  tnmo=0.0527704,1.54354,3.13984,4.97361 \
42  vnmo=1485.83,1861.25,2766.67,3318.75 \
43  tnmo=0.0527704,1.41161,3.50923,4.98681 \
44  vnmo=1507.92,1905.42,2943.33,3340.83 \
45  tnmo=0.0527704,1.20053,2.3219,3.0343,4.98681 \
46  vnmo=1485.83,1839.17,2722.5,3031.67,3407.08 \
```

15.4 NMO and Stack

Below is file Tnmo.sh. The output stack file size is 4.14 Mbytes.

```
1  #! /bin/sh
2  # File: Tnmo.sh
3  #       Apply NMO & stack
4  #  Input: line of CMPs
5  # Output: stacked seismic data
6
```

```
 7  # Set messages on
 8  set -x
 9
10  # User values
11   indata=Tdata/Tcmp5kgf.su
12  outdata=Tstack.su
13  myperc=95
14
15  # NMO
16  sunmo < $indata \
17  cdp=4175,4200,4225,4250,4275,4300,4325,4350,4375,4400,4425,
    4450,4475,4500,4525,4550,4575,4600,4625,4650,4675,4700 \
18  tnmo=0.0527704,0.527704,2.26913,3.219,3.7467,4.96042 \
19  vnmo=1485.83,1574.17,2325,2634.17,3252.5,4224.17 \
20  tnmo=0.0659631,1.0686,3.33773,5 \
21  vnmo=1463.75,1706.67,3053.75,3826.67 \
22  tnmo=0.0527704,0.844327,2.00528,3.37731,4.93404 \
23  vnmo=1507.92,1640.42,2325,3031.67,3426.41 \
24  tnmo=0.0395778,0.897098,2.41425,3.49604,4.97361 \
25  vnmo=1530,1640.42,2347.08,3009.58,4069.58 \
26  tnmo=0.0527704,0.620053,2.71768,3.64116,4.98681 \
27  vnmo=1463.75,1596.25,2479.58,2832.92,3583.75 \
28  tnmo=0.0395778,0.633245,2.62533,3.48285,4.98681 \
29  vnmo=1463.75,1596.25,2656.25,3208.33,3892.92 \
30  tnmo=0.0527704,0.633245,3.7467,4.98681 \
31  vnmo=1441.67,1552.08,3451.25,3848.75 \
32  tnmo=0.0527704,0.699208,2.8496,4.98681 \
33  vnmo=1507.92,1662.5,2567.92,3053.75 \
34  tnmo=0.0263852,1.04222,3.85224,4.97361 \
35  vnmo=1441.67,1772.92,3164.17,3716.25 \
36  tnmo=0.0527704,1.22691,3.10026,4.12929,5 \
37  vnmo=1463.75,1728.75,2590,3539.58,3804.58 \
38  tnmo=0.0527704,1.18734,2.4934,3.66755,5 \
39  vnmo=1463.75,1684.58,2413.33,2899.17,3252.5 \
40  tnmo=0.0527704,1.10818,2.94195,3.69393,4.97361 \
41  vnmo=1485.83,1706.67,2413.33,2921.25,3296.67 \
42  tnmo=0.0527704,0.91029,3.68074,4.98681 \
43  vnmo=1507.92,1772.92,3495.42,4091.67 \
44  tnmo=0.0659631,1.21372,3.69393,4.97361 \
45  vnmo=1530,1596.25,2965.42,3738.33 \
46  tnmo=0.0527704,1.21372,3.44327,5.01319 \
47  vnmo=1485.83,1662.5,2766.67,3694.17 \
48  tnmo=0.0395778,1.26649,3.79947,4.97361 \
49  vnmo=1463.75,1640.42,2877.08,3097.92 \
50  tnmo=0.0527704,2.24274,4.1161,4.97361 \
51  vnmo=1463.75,1883.33,2457.5,2943.33 \
52  tnmo=0.0395778,1.93931,3.25858,5.01319 \
53  vnmo=1441.67,1927.5,2457.5,3451.25 \
54  tnmo=0.0659631,1.27968,3.25858,4.97361 \
55  vnmo=1485.83,1618.33,2810.83,3826.67 \
56  tnmo=0.0527704,1.54354,3.13984,4.97361 \
57  vnmo=1485.83,1861.25,2766.67,3318.75 \
58  tnmo=0.0527704,1.41161,3.50923,4.98681 \
59  vnmo=1507.92,1905.42,2943.33,3340.83 \
60  tnmo=0.0527704,1.20053,2.3219,3.0343,4.98681 \
61  vnmo=1485.83,1839.17,2722.5,3031.67,3407.08 |
62
63  # Stack
64  sustack > $outdata
65
66  suximage < $outdata perc=$myperc label2=cdp key=cdp \
67          title="$outdata  perc=$myperc" &
68
```

```
69   # Exit politely from shell
70   exit
71
```

Script *iva* puts a continuation mark (\) at the end of all lines. At the end of line 61, we replaced the continuation mark with a pipe (|). After NMO, the data are piped to **sustack** (line 64).

Lines 66-67 make a plot using **suximage**. Remember, **suximage** reads the first *key* value (here it is *cdp*) and increments by "1" (see Section 7.7). If *key* values do not increment by "1" (for example, due to missing CMPs), the x-axis annotation will be wrong.

Figure 15.3 shows our stack of the Taiwan data.

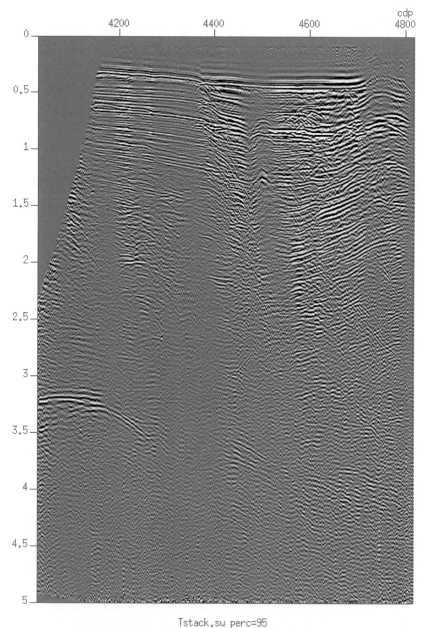

Figure 15.3: *Stack of Taiwan data.*

16. Taiwan: Migration

16.1 Constant Velocity Stolt Migration with Tmigcvp.sh

Below is the user area of *Tmigcvp.sh* for the Taiwan data.

- Remember, *cdpmin* and *cdpmax* refer to the range of CMPs that will have DMO applied to them. Previously, we decided that we will always let **sustolt** apply DMO to all stack traces (Section 9.2.1).

- The CDP bin distance, provided to us with the data files, is 16.667 m (parameter *dxcdp*, line 22).

- Because we do not know rms velocities for these data, we set *vscale=1.0* (line25). We process for image, not for velocities.

- Our first velocity is 1000 m/s and our last is 4000 m/s. Our velocity increment is 100 m/s. Here, we choose a large velocity range because we think the migration changes will be subtle.

```
11   #=================================================
12   # USER AREA -- SUPPLY VALUES
13   #-------------------------------------------------
14
15   # Seismic files
16   indata=Tdata/Tstack.su      # SU format
17   outdata=Tdata/Tmigcvp.su    # migration Constant Velocity Panels
18
19   # Migration variables
20   cdpmin=4027      # Start CDP value
21   cdpmax=4816      # End CDP value
22   dxcdp=16.667     # distance between adjacent CDP bins (m)
23   smig=1.0         # stretch factor (0.6 typical if vrms increasing)
24                    # [the "W" factor] (Default=1.0)
25   vscale=1.0       # scale factor to apply to velocities (Default=1.0)
26   lstaper=20       # length of side tapers (traces) (Default=0)
27   lbtaper=100      # length of bottom taper (samples) (Default=0)
28
29   # Velocity panel variables
30   firstv=1000      # first velocity value
31    lastv=4000      # last velocity value
32   increment=100    # velocity increment
33
34   numVtest=100     # use to limit number of velocity panels
35                    # otherwise, use very large value (100)
36
37   #=================================================
```

The output of *Tmigcvp.sh* is 31 constant velocity migration panels, 128 Mbytes.

16.2 Migration Movie

Below is script *Tmigmovie.sh* for the Taiwan data. We set *loop=1* (line 12) to run the movie forward continuously. Below, *fframe=1000* and *dframe=100* because we re-ran the migration (*Tmigcvp.sh*) for velocities 1000-4000 m/s at 100 m/s increment.

```
1   #! /bin/sh
2   # File: Tmigmovie.sh
3   #        Run a "movie" of the migration panels
4   #        Plot "title" shows panel velocity
5   #        Enter "xmovie" for mouse and keyboard options
```

```
 6
 7  set -x
 8
 9  indata=Tdata/Tmigcvp.su
10  perc=95
11
12  loop=1          # 1 = run panels forward continuously
13                  # 2 = run panels back and forth continuously
14                  # 0 = load all panels then stop
15
16  n1=1251         # number of time samples
17  d1=0.004        # time sample interval
18  n2=790          # number of traces per panel
19  d2=1            # trace spacing
20
21  width=550       # width of window
22  height=700      # height of window
23
24  fframe=1000     # velocity of first panel for title annotation
25  dframe=100      # panel velocity increment for title annotation
26
27  suxmovie < $indata perc=$perc loop=$loop \
28             n1=$n1 d1=$d1 n2=$n2 d2=$d2 \
29             width=$width height=$height \
30             fframe=$fframe dframe=$dframe \
31             title="Velocity %g" &
32
33  exit
34
```

Below is the **surange** output of *Tmigcvp.su*. Several facts from this output are useful for supplying values to *Tmigmovie.sh*. On line 21, you have to supply the number of time samples – the value of key *ns*. On line 17, you have to supply the time sample interval – the value of key *dt*.

```
        surange  <  Tmigcvp.su
24490 traces:
 tracl=(1,790)  tracr=(1,25344)  fldr=(1,31)  tracf=(1,144)  ep=(740,915)
 cdp=(4027,4816)  cdpt=(1,144)  trid=1 nvs=31 nhs=(1,48)
 offset=(1000,4000)  sdepth=80000 swdep=(2120000,3540000)  scalel=-10000
scalco=1
 sx=(-4773100,-3965250)  gx=(-4764600,-3600343)  gy=(2277,94805)  counit=3
tstat=12
 ns=1251 dt=4000 gain=9 afilf=160 afils=72
 lcf=3 hcf=160 lcs=6 hcs=72 year=95
 day=260 hour=(12,13)  minute=(0,59)  sec=(0,59)
```

The panel velocities are in the *offset* key and the number of velocities is in the *nvs* key. Using these two keys, you can calculate the values for *fframe* and *dframe* (if you do not want to check the *Tmigcvp.sh* values of *firstv* and *increment*).

Refer to Section 9.4 or the **xmovie** selfdoc for help using the mouse and keyboard with the movie. If you switch from Movie mode to Step mode by pressing the S key, you can step forward with the F key or backward with the B key. To restart the movie, press the C key. To quit, press the Q key.

Viewing the movie of single velocity migrations, it was difficult for us to pick a "best" migration. We think 2500 m/s is a reasonable image, but we are not happy with it.

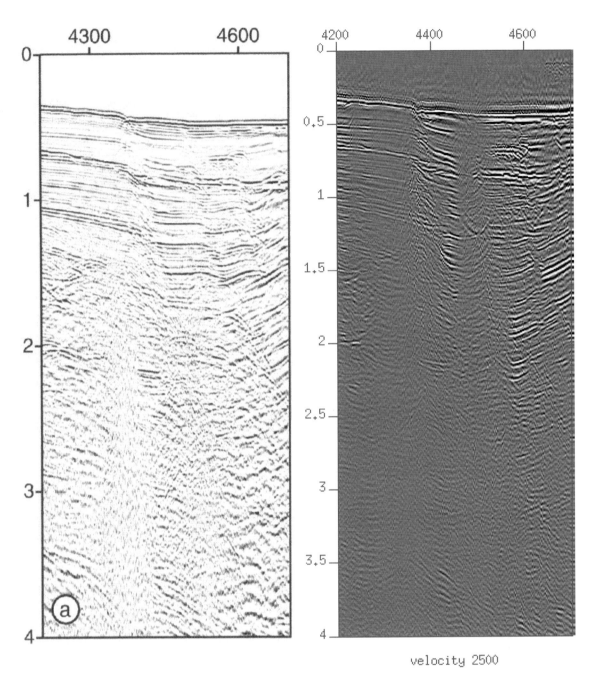

Figure 16.1: Left: *A portion of the 2-D line off the shelf area southwest of Taiwan. From Berndt and Moore (1999), Figure 6a (used with permission). Right: Our 2500 m/s constant velocity Stolt migration.*

Figure 16.1 puts our migration next to an image reported by Berndt and Moore (1999).

16.2 Discussion

While viewing the shot gathers, we did not mention that we saw sideswipe. Sideswipe is "evidence of a structural feature which lies off to the side of a line or traverse (Sheriff, 2002)." The problem with a 2-D survey is that real geology is 3-D. A 2-D survey is processed to image information from below the line of acquisition, but acoustic energy can reflect from side features as well as from "down."

Figure 16.2 shows a strong sideswipe event between 3.5-4 seconds. Other shot gathers show sideswipe at various times, generally at the same angle as in Figure 16.2 (for example, look carefully at shot 930).

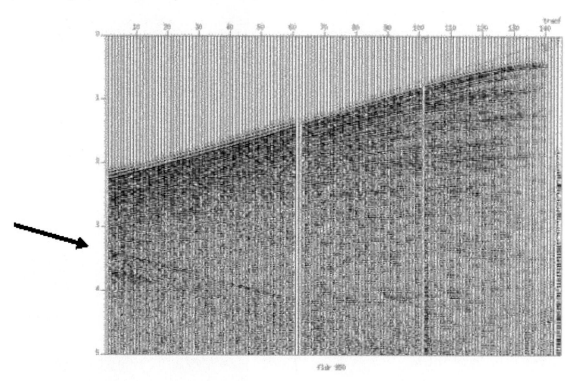

Figure 16.2: *Taiwan shot gather 950. The arrow points to sideswipe.*

When sideswipe (also called "out of plane reflections") occurs on 2-D data, we cannot "image" it because a 2-D line does not have enough information. We usually try various ways to eliminate sideswipe. One way is tau-p filtering. The tau-p filter is also called slant stack processing and radon filtering (Yilmaz 1987 and Yilmaz 2001).

Although we do not discuss tau-p filtering in this Primer, SU does have **suradon** and **sutaup**. Look at the examples in demos/Tau_P.

The paper by Berndt and Moore (1999) discusses several methods to attenuate sea-floor multiples, one of which they applied to this data set (their Figure 6). Two processes we recommend that you investigate are:

- prestack f-k or tau-p filtering and
- time-varying migration.

References

Berndt, C. & Moore, G.F., 1999, Dependence of multiple-attenuation techniques on the geologic setting: A case study from offshore Taiwan, The Leading Edge, Vol. 18, No. 1, 74-80.

Cohen, J.K. & Stockwell, Jr. J.W., 2002, CWP/SU: Seismic Unix Release No. 36: a free package for seismic research and processing, Center for Wave Phenomena, Colorado School of Mines.

Lamb, L., 1990, Learning the vi Editor, Sebastopol, California: O'Reilly & Associates, Inc..

Lindseth, R.O., 1982, Digital Processing of Geophysical Data – A Review, Tulsa: Society of Exploration Geophysicists (reprint).

Moore, G.F., Shipley, T.H., Stoffa, P.L., Karig, D.E., Taira, A., Kuramoto, S., Tokuyama, H., Suyehiro, K., 1990, Journal of Geopysical Research, Vol. 95, No. B6, 8753-8765.

Robinson, E.A., Treitel, S., 2000, Geophysical Signal Analysis (reprint), Tulsa: Society of Exploration Geophysicists, p. 278-279.

Sheriff, R.E., 2002, Encyclopedic Dictionary of Applied Geophysics, 4th Edition, Tulsa: Society of Exploration Geophysicists.

Stockwell, Jr. J.W., & Cohen, J.K., 2002, The New SU User's Manual, Golden: CWP, Colorado School of Mines.

Yilmaz, Ö., 1987, Seismic Data Processing, Tulsa: Society of Exploration Geophysicists.

Yilmaz, Ö., 2001, Seismic Data Analysis, Tulsa: Society of Exploration Geophysicists.

Appendix A: Seismic Un*x Legal Statement

Below, presented verbatim, is the Legal Statement that accompanies Seismic Un*x release 38.

This file is property of the Colorado School of Mines.

Copyright © 2004, Colorado School of Mines, all rights reserved.

Warranty Disclaimer:
NO GUARANTEES, OR WARRANTIES, EITHER EXPRESS OR IMPLIED, ARE PROVIDED BY CWP, CSM, ANY EMPLOYEE OR MEMBER OF THE AFORESAID ORGANIZATIONS, OR BY ANY CONTRIBUTOR TO THIS SOFTWARE PACKAGE, REGARDING THE ACCURACY, SAFETY, FITNESS FOR A PARTICULAR PURPOSE, OR ANY OTHER QUALITY OR CHARACTERISTIC OF THIS SOFTWARE, OR ANY DERIVATIVE SOFTWARE.

Export Restriction Disclaimer:
We believe that CWP/SU: Seismic Un*x is a low technology product that does not appear on the Department of Commerce CCL list of restricted exports. Accordingly, we believe that our product meets the qualifications of an ECCN (export control classification number) of EAR99 and we believe it fits the qualifications of NRR (no restrictions required), and is thus not subject to export restrictions of any variety.

Limited License:
The CWP/SU Seismic Un*x package (SU) is not public domain software, but it is available free under the following terms and conditions:

1. Permission to use, copy, and modify this software for any purpose without fee and within the guidelines set forth below is hereby granted, provided that the above copyright notice, the warranty disclaimer, and this permission notice appear in all copies, and the name of the Colorado School of Mines (CSM) not be used in advertising or publicity pertaining to this software without the specific, written permission of CSM.

2. The simple repackaging and selling of the SU package as is, as a commercial software product, is expressly forbidden without the prior written permission of CSM. Any approved repackaging arrangement will carry the following restriction: only a modest profit over reproduction costs may be realized by the reproducer.

Approved Reference Format:
In publications, please refer to SU as per the following example:

Cohen, J. K. and Stockwell, Jr. J. W., (200_), CWP/SU: Seismic Un*x Release No. __: a free package for seismic research and processing, Center for Wave Phenomena, Colorado School of Mines.

Recent articles about SU in peer-reviewed journals:
- Saeki, T., (1999), A guide to Seismic Un*x (SU)(2)---examples of data processing (part 1), data input and preparation of headers, Butsuri-Tansa (Geophysical Exploration), vol. 52, no. 5, 465-477.
- Stockwell, Jr. J. W. (1999), The CWP/SU: Seismic Un*x Package, Computers and Geosciences, May 1999.
- Stockwell, Jr. J. W. (1997), Free Software in Education: A case study of CWP/SU: Seismic Un*x, The Leading Edge, July 1997.
- Templeton, M. E., Gough, C.A., (1998), Web Seismic Un*x: Making seismic reflection processing more accessible, Computers and Geosciences, in press.

Acknowledgements:
SU stands for CWP/SU:Seismic Un*x, a processing line developed at Colorado School of Mines, partially based on Stanford Exploration Project (SEP) software.

Appendix B: Seismic Un*x at Michigan Tech

The use of SU on any given system depends on the "path" that is assigned by the system administrator. This Appendix is an example for system administrators who may want to emulate the approach we use at Michigan Technological University.

B.1   Using Seismic Un*x

At Michigan Tech, you must use SU within a special terminal (x-term) window. Do the following:

1. Log into a Unix machine.

2. Open an x-term window and enter **seismicx**. This opens another window, titled Seismic Un*x #, where # is your SU version number.

    The first time you do this, it will take a while, because you are creating a new subdirectory structure.

3. In the Seismic Un*x window, enter **pwd** (on some computers you have to use **cwd**) to see the directory in which you are now working; it should be in your home area, in a directory called **seismicx**.

4. Enter **ls** to see a listing of the files and subdirectories in that directory. You should see a subdirectory called **demos**.

5. Enter **cd demos** to change to that subdirectory, then enter **ls** to see a listing within it. When you want to move back up one level of directory, enter **cd ..** (Yes, that is: cd space dot dot. Note that if you enter **cd** by itself, you will go to your home directory.).

After you have worked in the **demos** directory for a while, if you decide you have made a mess of the **demos** directory, remove the directory and all its contents, then close the Seismic Un*x window. The next time you open a Seismic Un*x window, the system will create a fresh **demos** directory.

B.2   Terminal Windows

You must open a Seismic Un*x x-term window and work within it to execute SU commands. If you are not in a Seismic Un*x window, your SU commands will not work, even if you are in the **seismicx** directory (or any of its subdirectories). This is because the setup procedure that places the SU package in the user's "path" is only in place when the user is in an x-term window also created by the setup procedure.

C. Utility Programs

C.1 Introduction

Sometimes the library of Seismic Unix (SU) programs doesn't have exactly what we want. Sometimes it is just fun to write a program that does exactly what we want. This Appendix presents three C programs that do simple tasks in SU.

- **sukeycount**.c
- **sudumptrace**.c
- **tvnmoqc**.c

These three programs were written while we used Seismic Un*x version 38. You might find these programs in a later SU version. We are glad to contribute to the SU library.

Before we examine these programs, we recommend the programming sections of "The *New* SU User's Manual" (Stockwell & Cohen, 2002). The Manual web site is:

> http://www.cwp.mines.edu/sututor/sututor.html

The Manual's Contents web page is:

> http://www.cwp.mines.edu/sututor/node1.html

The Manual web page that describes shell programming, "Extending SU by shell programming," is:

> http://www.cwp.mines.edu/sututor/node128.html

This Appendix does not discuss shell programming. We think you have seen many shell examples in previous chapters. However, we recommend the above link because it helped us learn SU shell programming.

The SU web page that begins the section about C programming, "How to Write an SU Program," is:

> http://www.cwp.mines.edu/sututor/node136.html

Under this section, there are three web pages:

- "Setting up the Makefile" –

  > http://www.cwp.mines.edu/sututor/node137.html

- "A template SU program" –

  > http://www.cwp.mines.edu/sututor/node138.html

- "Writing a new program: `suvlength`" –

  > http://www.cwp.mines.edu/sututor/node139.html

You will understand the examples in this Appendix faster if you first study the Manual pages. Reading and understanding the Makefile web page will save you the effort of compiling and linking your program – the Makefile helps you compile and link.

In this Appendix, we do not teach C programming; we presume you are familiar with C. This Appendix is written to help you extend your knowledge of C into SU. However, we do not explain our programs in detail. We hope that our simple examples help you start on the road to becoming a C programmer in SU.

C.2 Program sukeycount

Below is the selfdoc of **sukeycount**: The user parameter of **sukeycount** is a key.

```
SUKEYCOUNT - sukeycount writes a count of a selected key

    sukeycount key=key1 < infile [> outfile]

Required parameters:
key=key1        One key word.

Optional parameters:
verbose=0   quiet
        =1   chatty

Writes the key and the count to the terminal or a text
   file when a change of key occurs. This does not provide
   a unique key count (see SUCOUNTKEY for that).
Note that for key values  1 2 3 4 2 5
   value 2 is counted once per occurrence since this program
   only recognizes a change of key, not total occurrence.

Examples:
    sukeycount < stdin key=fldr
    sukeycount < stdin key=fldr > out.txt
```

Program **sukeycount** reads the first trace. After that, each trace's key value is compared against the previous trace. If the new trace's key value is the same as the previous key value, a counter is incremented. The next trace is read and the comparison is made again. When the key value changes, **sukeycount** writes (to the screen or to a log file) the key name, the value that was recently tested, and the count of the traces with that value. After each write, the counter is re-set.

If the input data are a 2-D line sorted in CMP-order, a count of traces within CMPs is equivalent to a fold count. Suppose the first 16 traces of *cmp4.su* have the following *cdp* values:

sequential trace number	cdp value
1	1
2	2
3	3
4	3
5	4
6	4
7	5
8	5
9	5
10	6
11	6
12	6
13	7
14	7
15	7
16	7

Below, we use **sukeycount** on the CMP-ordered 2-D data.

```
sukeycount < cmp4.su key=cdp > model4Fold.txt
```

Below are the first 7 lines of output

```
cdp = 1        has  1  trace(s)
cdp = 2        has  1  trace(s)
cdp = 3        has  2  trace(s)
cdp = 4        has  2  trace(s)
cdp = 5        has  3  trace(s)
cdp = 6        has  3  trace(s)
cdp = 7        has  4  trace(s)
```

To generate Figure 7.2, we

- opened Excel,
- imported file *model4Fold.txt*,
- specified that the imported file is space-delimited, and
- used Excel's Chart Wizard to plot the two numeric columns.

Below is program **sukeycount**.c.

```
1   /* David Forel, Jan. 2005.*/
2
3   #include "su.h"
4   #include "segy.h"
5   #include "header.h"
6
7   /********************* self documentation *********************/
8   char *sdoc[] = {
9   "                                                            ",
10  " SUKEYCOUNT - sukeycount writes a count of a selected key   ",
11  "                                                            ",
12  "    sukeycount key=key1 < infile [> outfile]                ",
13  "                                                            ",
14  " Required parameters:                                       ",
15  " key=key1       One key word.                               ",
16  "                                                            ",
17  " Optional parameters:                                       ",
18  " verbose=0   quiet                                          ",
19  "         =1   chatty                                        ",
20  "                                                            ",
21  " Writes the key and the count to the terminal or a text     ",
22  "    file when a change of key occurs. This does not provide ",
23  "    a unique key count (see SUCOUNTKEY for that).           ",
24  " Note that for key values  1 2 3 4 2 5                      ",
25  "    value 2 is counted once per occurrence since this program ",
26  "    only recognizes a change of key, not total occurrence.  ",
27  "                                                            ",
28  " Examples:                                                  ",
29  "    sukeycount < stdin key=fldr                             ",
30  "    sukeycount < stdin key=fldr > out.txt                   ",
31  "                                                            ",
32  NULL};
33
34  /* Credits:
35   *
36   *   MTU: David Forel
37   */
38  /*************** end self doc ********************************/
39
40  segy tr ;
41
42  int
43  main(int argc, char **argv)
44  {
45      cwp_String key[SU_NKEYS] ;  /* array of keywords */
46      int nkeys ;                /* number of keywords to retrieve */
```

```
47      int iarg ;              /* arguments in argv loop */
48      int countkey = 0 ;      /* counter of keywords in argc loop */
49      cwp_String output ;     /* string representing output format */
50      int verbose = 0 ;       /* verbose ? */
51      cwp_String type1 ;      /* key string */
52      float sort, sortold ;   /* for comparing new/old key values */
53      int isort, isortold ;   /* for comparing new/old key values */
54      int gatherkount ;       /* counter of traces within key */
55      int itotal = 0 ;        /* counter of total traces */
56
57      /* Initialize */
58      initargs(argc, argv) ;
59      requestdoc(1) ;
60
61      sortold  = -99999. ;
62      isortold = -99999 ;
63
64      /* Get key values */
65      if (!getparint("verbose",&verbose)) verbose=0 ;
66      if ((nkeys=countparval("key"))!=0)
67      {
68          getparstringarray("key",key) ;
69      }
70      else
71      {
72          /* support old fashioned method for inputting key fields */
73          /* as single arguments:  sukeycount key1 */
74          if (argc==1) err("must set one key value!") ;
75
76          for (iarg = 1; iarg < argc; ++iarg)
77          {
78              cwp_String keyword ;  /* keyword */
79
80              keyword = argv[iarg] ;
81
82              if (verbose) warn("argv=%s",argv[iarg]);
83              /* get array of types and indexes to be set */
84              if ((strncmp(keyword,"output=",7)!=0))
85              {
86                  key[countkey] = keyword ;
87                  ++countkey ;
88              }
89              if (countkey==0) err("must set one key value!") ;
90              if (countkey>1) err("must set only one key value!") ;
91          }
92          nkeys=countkey;
93      }
94      if (nkeys>1) err("must set only one key value!") ;
95
96      printf("\n") ;
97
98      /* Loop over traces */
99      gatherkount = 0 ;
100     while (gettr(&tr))
101     {
102         /* Do not loop over keys because only one is used */
103         Value vsort ;
104         gethdval(&tr, key[0], &vsort) ;
105         type1 = hdtype(key[0]) ;
106
107         if (*type1 == 'f')  /* float header */
108         {
109             sort  = vtof(type1,vsort) ;
```

```
110
111                     /* Don't write just because first trace is new */
112                     if ( itotal == 0 ) sortold = sort ;
113
114                     if ( sort != sortold )
115                     {
116                         printf(" %8s = %f", key[0], sortold) ;
117                         printf("      has  %d  trace(s)", gatherkount) ;
118                         printf("\n") ;
119                         sortold = sort ;
120                         gatherkount = 0 ;
121                     }
122                     ++gatherkount ;
123                     ++itotal ;
124                 }
125             else  /* non-float header */
126             {
127                 isort = vtoi(type1,vsort) ;
128
129                 /* Don't write just because first trace is new */
130                 if ( itotal == 0 ) isortold = isort ;
131
132                 if ( isort != isortold )
133                 {
134                     printf(" %8s = %d", key[0], isortold) ;
135                     printf("      has  %d  trace(s)", gatherkount) ;
136                     printf("\n") ;
137                     isortold = isort ;
138                     gatherkount = 0 ;
139                 }
140                 ++gatherkount ;
141                 ++itotal ;
142             }
143         }
144
145     /* Write after last trace is read */
146     if (*type1 == 'f')
147     {
148         printf(" %8s = %f", key[0], sortold) ;
149         printf("      has  %d  trace(s)", gatherkount) ;
150         printf("\n") ;
151     }
152     else
153     {
154         printf(" %8s = %d", key[0], isortold) ;
155         printf("      has  %d  trace(s)", gatherkount) ;
156         printf("\n") ;
157     }
158     printf("\n       %d trace(s) read\n\n", itotal) ;
159
160     return(CWP_Exit()) ;
161 }
162
```

C.3   Program sudumptrace

A Seismic Un*x program to print trace amplitude values is **suascii**. (It also prints trace headers.) For each trace value, **suascii** prints two columns: the first column is a time count, the second column is the amplitude value. However, if the amplitude values for each trace were printed in separate columns, we could more easily compare the traces.

Below is the selfdoc of **sudumptrace**: The default value *num=4* means **sudumptrace** prints values for the first four traces it reads. No key names are required. If the user supplies key names, key names and their values are printed above each trace. If *hpf=1*, the print format of trace key values is exponential; the default format is floating point. Trace amplitudes are printed using exponential format.

```
SUDUMPTRACE - print selected header values and data.
              Print first num traces.
              Use SUWIND to skip traces.

sudumptrace < stdin [> ascii_file]

Optional parameters:
    num=4                     number of traces to dump
    key=key1,key2,...         key(s) to print above trace values
    hpf=0                     header print format is float
                              =1 print format is exponential


Examples:
  sudumptrace < inseis.su                PRINTS: 4 traces, no headers
  sudumptrace < inseis.su key=tracf,offset
  sudumptrace < inseis.su num=7 key=tracf,offset > info.txt
  sudumptrace < inseis.su num=7 key=tracf,offset hpf=1 > info.txt
```

In Section 12.3, we used **sudumptrace** to examine trace amplitudes of Taiwan shot gather (*fldr*) 930. We also used **suwind** to insure that the first traces into **sudumptrace** were the ones we wanted to examine.

Here, we use a small data set *mini3.su*. Below is the **surange** output of *mini3.su*.

```
        surange < mini3.su
7 traces:
 tracl=(129169,129175)  tracr=(14401,14407)  fldr=900 tracf=(1,7)  ep=840
 cdp=(4419,4425)  cdpt=(1,7)  trid=1 nhs=1 offset=(-3663,-3513)
 sdepth=80000 swdep=3110000 scalel=-10000 scalco=1 sx=-4450200
 gx=(-4110870,-4096381)  gy=(90923,94805)  counit=3 tstat=12 delrt=28
 ns=19 dt=4000 gain=9 afilf=160 afils=72
 lcf=3 hcf=160 lcs=6 hcs=72 year=95
 day=260 hour=12 minute=58 sec=40
```

File *mini3.su* has 7 traces and 19 time samples (*ns=19*). Also note that the delay time is 28 ms (*delrt=28*). Below, we use **sudumptrace** on *mini3.su*.

```
        sudumptrace < mini3.su key=tracl,cdp,tracf,ns

num traces = 4     num samples = 19

            tracl  129169.0000  129170.0000   129171.0000   129172.0000
              cdp    4419.0000    4420.0000     4421.0000     4422.0000
            tracf       1.0000       2.0000        3.0000        4.0000
               ns      19.0000      19.0000       19.0000       19.0000


Counter   Time    Values
      1   0.032   -4.6336e-01   1.5644e-01   -1.2573e-01    5.4677e-02
      2   0.036   -1.9764e+00   6.2943e-01   -5.3233e-01    2.4022e-01
      3   0.040   -4.0162e+00   1.1478e+00   -1.0636e+00    4.8795e-01
      4   0.044   -4.1991e+00   9.5269e-01   -1.0558e+00    4.4756e-01
      5   0.048   -3.1962e+00   5.0495e-01   -7.0221e-01    2.0370e-01
      6   0.052   -3.4293e+00   4.8322e-01   -6.9947e-01    2.3813e-01
      7   0.056   -3.3138e+00   2.8859e-01   -6.4975e-01    3.6629e-01
      8   0.060   -2.5544e+00   5.0591e-01   -4.7939e-01    3.4599e-01
```

9	0.064	-2.7929e+00	1.6010e+00	-5.5672e-01	4.0045e-01
10	0.068	-3.0077e+00	2.1449e+00	-5.4405e-01	4.9452e-01
11	0.072	-2.7386e+00	1.6686e+00	-5.5408e-01	5.8649e-01
12	0.076	-2.0205e+00	1.0067e+00	-7.8985e-01	7.6442e-01
13	0.080	-1.1539e+00	1.0423e+00	-9.8582e-01	9.8357e-01
14	0.084	-1.0390e+00	1.2934e+00	-1.0409e+00	1.1426e+00
15	0.088	-9.2735e-01	8.8366e-01	-8.3712e-01	1.2262e+00
16	0.092	-5.6792e-01	5.5448e-01	-5.3082e-01	1.3517e+00
17	0.096	-5.5705e-01	5.7637e-01	-5.0095e-01	1.4079e+00
18	0.100	-6.4979e-01	9.0827e-01	-5.6816e-01	1.4072e+00
19	0.104	-5.0397e-01	1.6227e+00	-3.2804e-01	1.5339e+00

The first output column is time sample count; the second column is time value (seconds). The next four columns are amplitude values of the first four input traces. The amplitude columns are unevenly spaced because they are tab-delimited.

The plot of *mini3.su*, Figure C.1, uses *key=tracf* to label the x-axis.

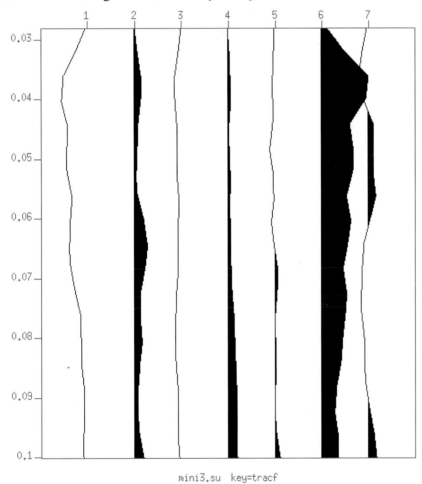

mini3.su key=tracf

Figure C.1: *Plot of mini3.su.*

For another test, we window *mini3.su* so the input to **sudumptrace** is the last 3 traces of the file.

```
suwind key=tracf min=5 < mini3.su | sudumptrace key=tracf,cdp,offset
```

Although we used the default *num=4*, only the last three traces were available for input to **sudumptrace**. Below is the screen output.

```
num traces = 3     num samples = 19
```

```
          tracf        5.0000        6.0000            7.0000
            cdp     4423.0000     4424.0000         4425.0000
         offset    -3563.0000    -3538.0000        -3513.0000
```

```
Counter     Time     Values
      1     0.032    -9.5147e-02    9.3358e-01    -2.3633e-01
      2     0.036    -3.1833e-01    3.8977e+00    -9.3778e-01
      3     0.040    -4.1830e-01    7.5830e+00    -1.5560e+00
      4     0.044    -2.4751e-01    7.1671e+00    -6.5207e-01
      5     0.048    -4.9481e-01    4.6742e+00     7.9582e-01
      6     0.052    -8.2958e-01    5.2009e+00     8.2085e-01
      7     0.056    -3.3455e-01    5.1407e+00     8.2723e-01
      8     0.060    -1.7911e-01    4.1310e+00     1.2068e+00
      9     0.064    -5.3361e-01    4.7825e+00     3.2759e-01
     10     0.068    -8.2213e-02    4.2018e+00    -7.1798e-01
     11     0.072     4.3474e-01    3.5753e+00    -9.7540e-01
     12     0.076     2.3809e-01    4.1400e+00    -1.1255e+00
     13     0.080     1.3812e-02    3.8605e+00    -9.5250e-01
     14     0.084     1.2431e-01    3.4150e+00    -5.5377e-01
     15     0.088     1.6596e-01    3.1037e+00    -5.0291e-01
     16     0.092     4.8231e-03    2.4580e+00    -3.4983e-01
     17     0.096    -2.8939e-02    2.1375e+00     2.9766e-01
     18     0.100     2.4795e-01    2.7438e+00     1.0379e+00
     19     0.104     7.8749e-01    2.6644e+00     1.3592e+00
```

Below is program **sudumptrace**.c.

```c
 1   /* David Forel    Jan 2005 */
 2
 3   #include "su.h"
 4   #include "segy.h"
 5   #include "header.h"
 6   #include <signal.h>
 7
 8   /********************* self documentation *********************/
 9   char *sdoc[] = {
10   "                                                            ",
11   " SUDUMPTRACE - print selected header values and data.       ",
12   "               Print first num traces.                      ",
13   "               Use SUWIND to skip traces.                   ",
14   "                                                            ",
15   " sudumptrace < stdin [> ascii_file]                         ",
16   "                                                            ",
17   " Optional parameters:                                       ",
18   "    num=4                     number of traces to dump      ",
19   "    key=key1,key2,...         key(s) to print above trace values ",
20   "    hpf=0                     header print format is float  ",
21   "                              =1 print format is exponential ",
22   "                                                            ",
23   " Examples:                                                  ",
24   "    sudumptrace < inseis.su         PRINTS: 4 traces, no headers ",
25   "    sudumptrace < inseis.su key=tracf,offset               ",
26   "    sudumptrace < inseis.su num=7 key=tracf,offset > info.txt    ",
27   "    sudumptrace < inseis.su num=7 key=tracf,offset hpf=1 > info.txt ",
28   "                                                            ",
29   NULL};
30
31   /* Credits:
32    *   MTU: David Forel
33    *
```

```
34      * Trace header field accessed: nt, dt, delrt
35      */
36     /*************** end self doc ********************************/
37
38     /* subroutine prototypes */
39     void dump(float *data, float dt, float *hedr, cwp_String *key,
40             float delrt, int nkeys, int ntr, int nt, int hpf) ;
41     static void closefiles(void) ;
42
43     /* Globals (so can trap signal) defining temporary disk files */
44     char tracefile[BUFSIZ] ;   /* filename for the file of traces */
45     char headerfile[BUFSIZ] ;  /* filename for the file of headers */
46     FILE *tracefp ;            /* fp for trace storage file */
47     FILE *headerfp ;           /* fp for header storage file */
48
49     segy tr ;
50
51     int
52     main(int argc, char **argv)
53     {
54         char *tmpdir ;                 /* directory path for tmp files */
55         cwp_Bool istmpdir=cwp_false ;  /* true for user given path */
56         float *hedr ;                  /* the headers */
57         float *data ;                  /* the data */
58
59         int nt ;                       /* number of trace samples */
60         float dt ;                     /* sample interval, sec */
61         float delrt ;                  /* delay recording time, sec */
62         cwp_String key[SU_NKEYS] ;     /* array of keywords */
63         cwp_String type ;              /* key string type */
64         int nkeys ;                    /* number of keywords */
65         int ikey, itr, ntr = 0 ;       /* counters */
66         int num ;                      /* number of traces to dump */
67         int numtr = 4 ;                /* number of traces to dump */
68         int hpf ;                      /* header print format */
69
70         /* Initialize */
71         initargs(argc, argv) ;
72         requestdoc(1) ;
73
74         /* Look for user-supplied tmpdir */
75         if (!getparstring("tmpdir",&tmpdir) &&
76             !(tmpdir = getenv("CWP_TMPDIR"))) tmpdir="";
77         if (!STREQ(tmpdir, "") && access(tmpdir, WRITE_OK))
78             err("you can't write in %s (or it doesn't exist)", tmpdir);
79
80         /* Get values from first trace */
81         if (!gettr(&tr)) err("can't get first trace");
82         nt = (int) tr.ns ;               /* Get nt */
83         dt = ((double) tr.dt)/1000000.0 ;   /* microsecs to secs */
84         if (!dt) getparfloat("dt", &dt) ;
85         if (!dt) MUSTGETPARFLOAT("dt", &dt) ;
86         delrt = ((double) tr.delrt)/1000.0 ; /* millisecs to secs */
87
88         /* Get parameters */
89         if (getparint ("num", &num)) numtr = num ;
90         if ((nkeys=countparval("key"))!=0) getparstringarray("key",key) ;
91         hedr = ealloc1float(nkeys*numtr) ;  /* make space for headers */
92         if (!getparint ("hpf", &hpf)) hpf = 0 ;
93
94         /* Store traces, headers in tempfiles */
95         if (STREQ(tmpdir,""))
96         {
```

```
97          tracefp = etmpfile();
98          headerfp = etmpfile();
99
100         do
101         {
102             ++ntr;
103             efwrite(&tr, HDRBYTES, 1, headerfp);
104             efwrite(tr.data, FSIZE, nt, tracefp);
105
106             /* Get header values */
107             for (ikey=0; ikey<nkeys; ++ikey)
108             {
109                 Value val;
110                 float fval;
111
112                 gethdval(&tr, key[ikey], &val) ;
113                 type = hdtype(key[ikey]) ;
114                 fval = vtof(type,val) ;
115                 hedr[(ntr-1)*nkeys+ikey] = fval ;
116             }
117
118         }
119         while (ntr<numtr && gettr(&tr)) ;
120
121     }
122     else  /* user-supplied tmpdir */
123     {
124         char directory[BUFSIZ];
125         strcpy(directory, tmpdir);
126         strcpy(tracefile, temporary_filename(directory));
127         strcpy(headerfile, temporary_filename(directory));
128         /* Handle user interrupts */
129         signal(SIGINT, (void (*) (int)) closefiles);
130         signal(SIGQUIT, (void (*) (int)) closefiles);
131         signal(SIGHUP,  (void (*) (int)) closefiles);
132         signal(SIGTERM, (void (*) (int)) closefiles);
133         tracefp = efopen(tracefile, "w+");
134         headerfp = efopen(headerfile, "w+");
135         istmpdir=cwp_true;
136
137         do
138         {
139             ++ntr;
140             efwrite(&tr, HDRBYTES, 1, headerfp);
141             efwrite(tr.data, FSIZE, nt, tracefp);
142
143             /* Get header values */
144             for (ikey=0; ikey<nkeys; ++ikey)
145             {
146                 Value val;
147                 float fval;
148
149                 gethdval(&tr, key[ikey], &val) ;
150                 type = hdtype(key[ikey]) ;
151                 fval = vtof(type,val) ;
152                 hedr[(ntr-1)*nkeys+ikey] = fval ;
153             }
154
155         }
156         while (ntr<numtr && gettr(&tr)) ;
157
158     }
159
```

```
160        /* Rewind after read, allocate space */
161        erewind(tracefp);
162        erewind(headerfp);
163        data = ealloc1float(nt*ntr);
164
165        /* Load traces into data and close tmpfile */
166        efread(data, FSIZE, nt*ntr, tracefp);
167        efclose(tracefp);
168        if (istmpdir) eremove(tracefile);
169
170        rewind(headerfp);
171        rewind(tracefp);
172
173        /* Do trace work */
174        dump(data, dt, hedr, key, delrt, nkeys, ntr, nt, hpf) ;
175
176        /* close */
177        efclose(headerfp);
178        if (istmpdir) eremove(headerfile);
179
180        free1(hedr) ;
181        free1(data) ;
182
183        return(CWP_Exit()) ;
184    }
185
186
187    void dump(float *data, float dt, float *hedr, cwp_String *key,
188              float delrt, int nkeys, int ntr, int ntime, int hpf)
189    /*
190      Dump headers and traces in column format; one trace per column.
191      INPUT:
192        data    array of trace data values
193        dt      trace sample interval
194        delrt   delay recording time, sec
195        hedr    array of trace headers
196        key     array of key names
197        nkeys   number of keys (headers) to print above trace values
198        ntr     number of traces to dump
199        ntime   number of time samples on trace
200        hpf     header print format flag: 0=float, 1=exponential
201      OUTPUT: none
202    */
203    {
204        int i, ikey, j, k, m ;              /* counters */
205
206        printf("\nnum traces = %d    num samples = %d \n\n",ntr,ntime) ;
207
208        for (k=0; k<nkeys; ++k)                    /* Print headers */
209        {
210            printf("%17s  ", key[k]) ;
211            for (m=0; m<ntr; ++m)
212            {
213                if (hpf==0)
214                {
215                    printf("%11.4f\t", hedr[m*nkeys + k]) ;
216                }
217                else
218                {
219                    printf("%11.4e\t", hedr[m*nkeys + k]) ;
220                }
221            }
222            putchar('\n') ;
```

```
223      }
224
225      putchar('\n') ;
226
227      printf("\nCounter    Time      Values\n") ;   /* Column titles */
228
229      for (i=1; i<=ntime; ++i)                      /* Print trace values */
230      {
231          printf(" %6d ", i) ;
232          printf(" %8.3f   ", dt*(i)+delrt) ;
233          for (j=1; j<=ntr; ++j)
234          {
235              printf("%11.4e\t", data[(j-1)*ntime+(i-1)]) ;
236          }
237          putchar('\n') ;
238      }
239      putchar('\n') ;
240
241  }
242
243
244  /* for graceful interrupt termination */
245  static void closefiles(void)
246  {
247      efclose(headerfp);
248      efclose(tracefp);
249      eremove(headerfile);
250      eremove(tracefile);
251      exit(EXIT_FAILURE);
252  }
253
```

C.4   Program *tvnmoqc*

Program **suximage** can put lines on the image when the x-y line values are available as two columns. The input to **sunmo**, a file of CDP values and associated row pairs of *tnmo-vnmo* values (see Section 8.2) can provide values for an **suximage** line overlay. The task is to overlay a CDP's time-velocity values on the **suximage** velan. The difficulty is that the **sunmo** rows must be converted to columns for **suximage**.

Using *mode=2*, **tvnmoqc** converts a CDP's corresponding tnmo-vnmo rows to columns. The output file is named by a prefix that is supplied by the user (required parameter *prefix* when *mode=2*); the output file name suffix is the CDP number. For example, in line 15 of the code (below), if the user supplies *prefix=velqc*, the first output file is named *velqc.15* and the second output file is named *velqc.35*.

In either *mode* 1 or 2, **tvnmoqc** also makes several checks that **sunmo** does – most of the parameter checks were taken from **sunmo**.

- CDP values must be supplied.

- There must be the same number of *tnmo* and *vnmo* series as there are CDPs.

- And several other checks that **sunmo** does.

One important difference between **tvnmoqc** and **sunmo** is that **tvnmoqc** does not stop if a time series does not always increase. If a time series pair is found that does not increase, a message is printed to the screen and processing continues. You can use **tvnmoqc** as a quality check of the *cdp-tnmo-vnmo* values before using **sunmo**.

Section 8.2 shows examples of using *mode=1* and *mode=2*.

In **tvnmoqc**.c:

Lines 99-139 were taken almost line-by-line from **sunmo**. These lines check that the cdp-tnmo-vnmo values have acceptable values for **sunmo**.

Lines 137-138 are the diagnostic print to the screen.

Line 143 creates the output file name.

Line 144 calls function **mktvfile** that (1) converts the tnmo-vnmo values from rows to columns and (2) writes the t-v columns to the file.

```
 1   /* David Forel    May 2005 */
 2
 3   #include "par.h"
 4   #include <string.h>
 5
 6   /********************** self documentation ************************/
 7   char *sdoc[] = {
 8   "                                                            ",
 9   " TVNMOQC - Check tnmo-vnmo pairs; create t-v column files   ",
10   "                                                            ",
11   " tvnmoqc [parameters] cdp=... tnmo=... vnmo=...             ",
12   "                                                            ",
13   "    Example:                                                ",
14   " tvnmoqc mode=1 \\                                          ",
15   " cdp=15,35 \\                                               ",
16   " tnmo=0.0091,0.2501,0.5001,0.7501,0.9941 \\                 ",
17   " vnmo=1497.0,2000.0,2500.0,3000.0,3500.0 \\                 ",
18   " tnmo=0.0082,0.2402,0.4902,0.7402,0.9842 \\                 ",
19   " vnmo=1495.0,1900.0,2400.0,2900.0,3400.0                   ",
20   "                                                            ",
21   " Required Parameter:                                        ",
22   "    prefix=          Prefix of output t-v file(s)           ",
23   "                     Required only for mode=2               ",
24   "                                                            ",
25   " Optional Parameter:                                        ",
26   "    mode=1           1=qc: check that tnmo values increase  ",
27   "                     2=qc and output t-v files             ",
28   "                                                            ",
29   " mode=1                                                     ",
30   "    TVNMOQC checks that there is a tnmo and vnmo series for each CDP ",
31   "     and it checks that each tnmo series increases in time. ",
32   "                                                            ",
33   " mode=2                                                     ",
34   "    TVNMOQC does mode=1 checking, plus ...                  ",
35   "                                                            ",
36   "    TVNMOQC converts par (MKPARFILE) values written as:     ",
37   "                                                            ",
38   "            cdp=15,35,...,95 \\                             ",
39   "            tnmo=t151,t152,...,t15n \\                      ",
40   "            vnmo=v151,v152,...,v15n \\                      ",
41   "            tnmo=t351,t352,...,t35n \\                      ",
42   "            vnmo=v351,v352,...,v35n \\                      ",
43   "            tnmo=... \\                                     ",
44   "            vnmo=... \\                                     ",
45   "            tnmo=t951,t952,...,t95n \\                      ",
46   "            vnmo=v951,v952,...,v95n \\                      ",
47   "                                                            ",
48   "    to column format. The format of each output file is:    ",
49   "                                                            ",
50   "            t1 v1                                           ",
51   "            t2 v2                                           ",
52   "            ...                                             ",
```

```
53   "          tn vn                                                    ",
54   "                                                                   ",
55   "   One file is output for each input pair of tnmo-vnmo series.     ",
56   "                                                                   ",
57   "   A CDP VALUE MUST BE SUPPLIED FOR EACH TNMO-VNMO ROW PAIR.        ",
58   "                                                                   ",
59   "   Prefix of each output file is the user-supplied value of        ",
60   "      parameter PREFIX.                                            ",
61   "   Suffix of each output file is the cdp value.                    ",
62   "   For the example above, output files names are:                 ",
63   "      PREFIX.15  PREFIX.35  ...  PREFIX.95                         ",
64   "                                                                   ",
65   NULL};
66
67   /* Credits:
68    *      MTU: David Forel (adapted from SUNMO)
69    */
70   /*************** end self doc ****************************************/
71
72   void mktvfile(char outfile[], int ntnmo, float *tnmo, float *vnmo);
73   int main(int argc, char **argv)
74   {
75      int k;              /* index used in loop */
76      int mode;           /* mode=1: qc; mode=2: qc + make cdp-t-v file */
77      int icdp;           /* index into cdp array */
78      int ncdp;           /* number of cdps specified */
79      int *cdp;           /* array[ncdp] of cdps */
80      int nvnmo;          /* number of vnmos specified */
81      float *vnmo;        /* array[nvnmo] of vnmos */
82      int ntnmo;          /* number of tnmos specified */
83      float *tnmo;        /* array[ntnmo] of tnmos */
84      cwp_String prefix;  /* prefix of output files */
85      char dot[] = ".";   /* for output file name */
86      char outfile[80];   /* output file name */
87
88      /* Hook up getpar */
89      initargs(argc, argv);
90      requestdoc(1);
91
92      /* Get parameters */
93      if(!getparint("mode",&mode))mode=1;
94      if(mode==2)
95         if (!getparstring("prefix", &prefix))
96            err("When mode=2, you must supply a prefix name.");
97
98      /* Are there cdp values and vnmo-tnmo sets for each cdp? */
99      ncdp = countparval("cdp");
100     warn("This file has %i CDPs.",ncdp);
101     if (ncdp>0) {
102        if (countparname("vnmo")!=ncdp)
103           err("A vnmo set must be specified for each cdp");
104        if (countparname("tnmo")!=ncdp)
105           err("A tnmo set must be specified for each cdp");
106     } else {
107        err("A cdp value must be supplied for each tnmo-vnmo set");
108     }
109
110     /* Get cdp values */
111     cdp = ealloc1int(ncdp);
112     if (!getparint("cdp",cdp))
113        err("A cdp value must be supplied for each tnmo-vnmo set");
114
115     /* Get tnmo-vnmo values */
```

```
116    for (icdp=0; icdp<ncdp; ++icdp)
117    {
118        nvnmo = countnparval(icdp+1,"vnmo");
119        ntnmo = countnparval(icdp+1,"tnmo");
120        if (nvnmo!=ntnmo)
121            err("number of vnmo and tnmo values must be equal");
122        if (nvnmo==0) err("Each cdp must have at least one velocity");
123        if (ntnmo==0) err("Each cdp must have at least one time");
124
125        vnmo = ealloc1float(nvnmo);
126        tnmo = ealloc1float(nvnmo);
127
128        if (!getnparfloat(icdp+1,"vnmo",vnmo))
129            err("Each cdp must have at least one velocity");
130        if (!getnparfloat(icdp+1,"tnmo",tnmo))
131            err("Each cdp must have at least one time");
132        for (k=1; k<ntnmo; ++k)
133            if (tnmo[k]<=tnmo[k-1]) {
134                warn("tnmo values must increase for use in NMO");
135                /* A tnmo series that does not increase is not an
136                    error in velocity QC (but IS an error in SUNMO) */
137                warn("For cdp=%i, check times %g and %g.",
138                        cdp[icdp],tnmo[k-1],tnmo[k]);
139            }
140
141        /* write cdp-tnmo-vnmo values to file "prefix" "dot" "cdp" */
142        if (mode == 2) {
143            sprintf(outfile,"%s%s%i",prefix,dot,cdp[icdp]);
144            mktvfile(outfile,ntnmo,tnmo,vnmo);
145        }
146
147        free1float(vnmo);
148        free1float(tnmo);
149    }
150
151    free1int(cdp);
152    warn("End of cdp-tnmo-vnmo check.");
153
154    return(CWP_Exit());
155 }
156
157
158 void mktvfile(char outfile[], int ntnmo, float *tnmo, float *vnmo)
159 {
160    int i;         /* index used in loop */
161    FILE *fptr;   /* file pointer for output */
162
163    fptr = efopen(outfile, "w");
164    for (i=0; i<ntnmo; ++i) {
165        fprintf(fptr,"%g   %g\n",tnmo[i],vnmo[i]);
166    }
167    efclose(fptr);
168 }
169
```

C.5 *Conclusion*

As the SU User's Manual states, "The secret to efficient SU coding is finding an existing program similar to the one you want to write." For example, if you want part of your program to modify trace headers (keys), study **suchw**.c. If you want to modify trace amplitudes in a single-trace process (Section 10.6), we suggest you study **sugain**.c.

D. Makefiles: Alternative to Shell Scripts
by Chris Liner

D.1 Introduction

Makefiles are commonly used for software installation on Unix systems. But **make** can be used for other purposes as well. As a small example, I will describe a makefile (found in Section D.4) that reproduces the functionality of the shell script *model1.sh* (Section 4.2).

First, some general comments about working with makefiles. There can be only one makefile in a directory and, for our purposes, it will always be named "makefile." It is a text file like any other and can therefore be edited by whatever tools you use to edit shell scripts. I use **vi**, which is a simple editor built into every Unix system. To speed things up. I define an alias in my environment called **vimake**:

```
alias vimake 'vi makefile'
```

There are variations from system to system, but this kind of alias is usually defined in a "dot" file, such as .login or .cshrc (talk to your local system expert for advice). With this alias in place, you can navigate to the working directory you want, then type **vimake** and be editing the makefile.

A makefile can perform a simple or complex series of interrelated tasks. For example, the Stanford Exploration Project at Stanford University uses **make** to generate processing results, figures for publication, and even entire books. To be specific, at this time they use a variant called **gmake**, but I will describe generic **make** here.

D.2 Discuss the Makefile

Before describing what is inside the makefile, let's just run it. In what follows, % is the Unix prompt. The first thing to try is:

```
% make
```

In a well-designed makefile, this will do nothing except echo some information. The last thing you want is a dangerous makefile sitting there waiting for you to come back in 3 months and casually enter **make** only to have it remove or modify files that cannot be recovered. Enough on safe computing.

For our makefile, typing:

```
% make
```

invokes a default response which is just a little text explaining what is available via this makefile:

```
Available Makes:
  model1.dat ps clean
  Example: make model1.dat
```

This tells us we have three **make** choices. The first creates model data, the second creates a postscript file showing the model, and the third cleans the current working directory.

Let's look inside the makefile now. The first things we see are some parameters that will be used by later **make** items. Anything with a # in the first location is a comment ignored by **make**, but very useful for anyone who wants to understand your makefile (including you at some later date). So the lines:

```
# Experiment Number
num=1
```

give a comment and then a value is assigned to the parameter named *num*. Note that in the actual makefile these lines are flush left. Makefiles are touchy about alignments and spaces versus tabs, so it is always a good idea to copy a working makefile and then modify it to your needs; otherwise, a lot of time can be wasted trying to figure out formatting problems. Parameter *num* now has the value 1, and to use it somewhere later we need to refer to it in a particular way. For example we can use $(num) or ${num}; for our purposes these are equivalent. The next four lines of the makefile include two more comments and two parameter assignments:

```
# Name output binary model file
modfile=model${num}.dat
# Name output encapsulated Postscript image file
psfile=model${num}.eps
```

Notice that, as in shell scripts, the parameter assignments can be embedded, meaning that parameters can be set based on previously set parameters. Parameters *modfile* and *psfile* are thus defined in terms of the *num* parameter.

So much for parameter assignments; now on to the main action. The default action of the makefile is described by:

```
default:
        @echo " "
        @echo "Available Makes:"
        @echo "  model${num}.dat ps clean"
        @echo "  Example: make model${num}.dat"
        @echo " "
```

Again, it is very important that "default:" begin in column 1 and end with a colon, and equally important that all lines in the default routine begin with a tab, not a sequence of spaces. The makefile interprets these lines in this way: "When the user types make with no following characters, I will go to the first executable step, which happens to be named *default* in this case, and it tells me to echo 5 lines of text back to the command line, then stop."

When we enter **make**, we again see:

```
Available Makes:
  model1.dat ps clean
  Example: make model1.dat
```

verifying that parameters like ${*num*} have been replaced with their values, thus generating an echo like *model1.dat* from *model${num}.dat*.

The next **make** item actually gets something done:

```
${modfile}:
        rm -f ${psfile}
        trimodel xmin=0 xmax=6 zmin=0 zmax=2 \
          xedge=0,6 \
          .
          .
          .

        sfill=0.1,1.5,0,0,0.25,0,0 \
```

```
      > ${modfile}
```

The name of this item is given by parameter *modfile*, which in turn depends on parameter *num*. In fact, it is not easy to tell what we type to invoke this item. But from earlier typing of **make**, we know it is:

```
% make model1.dat
```

that will get the job done.

The first action is to remove the postscript file whose name is stored in parameter *psfile*. Next, SU program **trimodel** is executed with several lines of parameters (not all are shown here), and the resulting model file is created with a name given by parameter *modfile*. As with SU shell scripts, each continuation line must end in a "\",not a space. To get secondary indentation (good style), use a tab and 2 spaces. The number of spaces is up to you, but the tab is necessary or the makefile will not work.

The next item in our makefile is invoked by:

```
% make ps
```

This executes the following **make** code:

```
ps: ${modfile}
        # Create a Postscipt file of the model
        #   Set gtri=1.0 to see sloth triangle edges
        spsplot < ${modfile} \
          gedge=0.5 gtri=2.0 gmin=0 gmax=1 \
          title="Earth Model - 5 layers   [M${num}]" \
          labelz="Depth (km)" labelx="Distance (km)" \
          dxnum=1.0 dznum=0.5 wbox=6 hbox=2 \
        > ${psfile} \
```

Note the name of this item is hardwired to "ps" and that something sits on the name line to the right of the colon (:). **Make** interprets this line this way: "To **make** ps, I need a file called *model1.dat*. If it exists in the current directory, then I will proceede to execute this item. If it does not exist, then I will **make** it before continuing." This is an example of a conditional **make**. To **make** ps, we need to **make** *model1.dat*, unless a file of that name is already in the directory. This is a powerful idea. For example, a finite difference simulation might take 24 hours to create data that we want to display. With a conditional **make** item, the simulation will not be redone unless the data file is missing. If the data file is there, then only the display will be done.

For this to work right, **make** items have to be strictly matched to their dependencies. Consider the following example **make** items:

```
data:
        suspike ... > data.su
ps: data
        supsimage .... > fig.ps
```

If we enter "**make ps**", then **make** thinks it needs a file named "data." Not finding it, **make** will run the **make** item "data" and generate a *data.su* file. There is still no file named "data," but **make** is happy it did something logical and it goes on to create the image. So this is not a conditional **make**, but an absolute one. Invoking "**make ps**" will always make new data. The way to make this conditional is:

```
data.su:
        suspike ... > data.su
ps: data.su
```

```
supsimage .... > fig.ps
```

Notice that the name of the "data.su" **make** item is matched to the file name it creates, and both are matched to the file need of the "ps" item. Now if we invoke "**make ps**" it will only remake *data.su* if it is not present in the local directory.

D.3 "clean" and Final Thoughts

The last **make** item has no dependencies. It simply removes unwanted files.

```
clean:
        rm *.dat *.eps
```

It is good practice to design makefiles so that "**make clean**" leaves only the makefile and necessary files in the directory. By "necessary," I mean permanent data files (like a SEG-Y archived version) and that is about it. Everything else should be generated by the makefile. Also, if you are working on a publication, **make** items can be keyed to your document. For example, "**make fig11**" might generate fig11 from scratch. This goes a long way toward the goal of reproducible research, and also lets other people quickly adapt your work to their own data and problems.

D.4 The Makefile

```
# Experiment Number
num=1
# Name output binary model file
modfile=model${num}.dat
# Name output encapsulated Postscript image file
psfile=model${num}.eps

default:
        @echo " "
        @echo "Available Makes:"
        @echo "   model${num}.dat ps clean"
        @echo "   Example: make model${num}.dat"
        @echo " "

${modfile}:
        rm -f ${psfile}
        trimodel xmin=0 xmax=6 zmin=0 zmax=2 \
          xedge=0,6 \
          zedge=0,0 \
          sedge=0,0 \
          xedge=0,2,4,6 \
          zedge=0.30,0.50,0.20,0.30 \
          sedge=0,0,0,0 \
          xedge=0,2,4,6 \
          zedge=0.55,0.75,0.45,0.55 \
          sedge=0,0,0,0 \
          xedge=0,2,4,6 \
          zedge=0.65,0.85,0.55,0.65 \
          sedge=0,0,0,0 \
          xedge=0,2,4,6 \
          zedge=1.30,1.30,1.60,1.20 \
          sedge=0,0,0,0 \
          xedge=0,6 \
          zedge=2,2 \
```

```
            sedge=0,0 \
            kedge=1,2,3,4,5,6 \
            sfill=0.1,0.1,0,0,0.44,0,0 \
            sfill=0.1,0.4,0,0,0.40,0,0 \
            sfill=0.1,0.6,0,0,0.35,0,0 \
            sfill=0.1,1.0,0,0,0.30,0,0 \
            sfill=0.1,1.5,0,0,0.25,0,0 \
        > ${modfile}

ps: ${modfile}
        # Create a Postscipt file of the model
        #   Set gtri=1.0 to see sloth triangle edges
        spsplot < ${modfile} \
            gedge=0.5 gtri=2.0 gmin=0 gmax=1 \
            title="Earth Model - 5 layers  [M${num}]" \
            labelz="Depth (km)" labelx="Distance (km)" \
            dxnum=1.0 dznum=0.5 wbox=6 hbox=2 \
        > ${psfile} \

clean:
        rm *.dat *.eps
```

D.5 *The Makefile Edited*

Unfortunately, we cannot usually see tabs, just their effect on spacing. Since tabs are an essential part of makefiles, the makefile of Section D.4 is presented again below. This time, invisible tabs are replaced by <TAB> and line numbers are added for discussion.

Notice the unexpected tab on line 15. Do not overlook the ">" on lines 42 and 52.

```
 1    # Experiment Number
 2    num=1
 3    # Name output binary model file
 4    modfile=model${num}.dat
 5    # Name output encapsulated Postscript image file
 6    psfile=model${num}.eps
 7
 8    default:
 9    <TAB>@echo " "
10    <TAB>@echo "Available Makes:"
11    <TAB>@echo "  model${num}.dat ps clean"
12    <TAB>@echo "  Example: make model${num}.dat"
13    <TAB>@echo " "
14
15    ${modfile}:<TAB>
16    <TAB>rm -f ${psfile}
17    <TAB>trimodel xmin=0 xmax=6 zmin=0 zmax=2 \
18    <TAB> xedge=0,6 \
19    <TAB> zedge=0,0 \
20    <TAB> sedge=0,0 \
21    <TAB> xedge=0,2,4,6 \
22    <TAB> zedge=0.30,0.50,0.20,0.30 \
23    <TAB> sedge=0,0,0,0 \
24    <TAB> xedge=0,2,4,6 \
25    <TAB> zedge=0.55,0.75,0.45,0.55 \
26    <TAB> sedge=0,0,0,0 \
27    <TAB> xedge=0,2,4,6 \
```

```
28    <TAB>   zedge=0.65,0.85,0.55,0.65 \
29    <TAB>   sedge=0,0,0,0 \
30    <TAB>   xedge=0,2,4,6 \
31    <TAB>   zedge=1.30,1.30,1.60,1.20 \
32    <TAB>   sedge=0,0,0,0 \
33    <TAB>   xedge=0,6 \
34    <TAB>   zedge=2,2 \
35    <TAB>   sedge=0,0 \
36    <TAB>   kedge=1,2,3,4,5,6 \
37    <TAB>   sfill=0.1,0.1,0,0,0.44,0,0 \
38    <TAB>   sfill=0.1,0.4,0,0,0.40,0,0 \
39    <TAB>   sfill=0.1,0.6,0,0,0.35,0,0 \
40    <TAB>   sfill=0.1,1.0,0,0,0.30,0,0 \
41    <TAB>   sfill=0.1,1.5,0,0,0.25,0,0 \
42    <TAB>> ${modfile}
43
44    ps: ${modfile}
45    <TAB># Create a Postscipt file of the model
46    <TAB>#   Set gtri=1.0 to see sloth triangle edges
47    <TAB>spsplot < ${modfile} \
48    <TAB>   gedge=0.5 gtri=2.0 gmin=0 gmax=1 \
49    <TAB>   title="Earth Model - 5 layers   [M${num}]" \
50    <TAB>   labelz="Depth (km)" labelx="Distance (km)" \
51    <TAB>   dxnum=1.0 dznum=0.5 wbox=6 hbox=2 \
52    <TAB>> ${psfile} \
53
54    clean:
55    <TAB>rm *.dat *.eps
```

Appendix E: On the CDs

E.1 Category List

The following scripts are included:

Ngazmig.sh	fxdisp.scr	killer2.sh	psmerge2a.sh
Niviewcvp.scr	fxdisp.sh	migStolt.sh	psmerge2b.sh
Niviewcvp.sh	idecon.scr	migcvp.sh	psmerge3.sh
Nmigcvp.sh	idecon.sh	migmovie.sh	shotrwd1.sh
Nmigmovie.sh	ifk.scr	model1.sh	shotrwd2.sh
Nnmo.sh	ifk.sh	model2.sh	shotrwd3.sh
Nsort2cmp.sh	igain.scr	model3.sh	shotrwd4.sh
Tmigcvp.sh	igain.sh	model4.sh	showcmp.sh
Tmigmovie.sh	iva.scr	myplot.sh	showshot.sh
Tnmo.sh	iva.sh	nmo4.sh	showshotB.sh
acq1.sh	iview.scr	nmo4a.sh	sort2cmp.sh
acq2.sh	iview.sh	oz14prep.sh	tf.sh
acq3.sh	iviewcvp.scr	oz14velan.sh	tvQC.sh
acq4.sh	iviewcvp.sh	psmerge1a.sh	velanQC.scr
acq4shot.sh	killer.sh	psmerge1b.sh	velanQC.sh

The following ASCII file is included:

killer.txt (output of killer.sh)

The following seismic data files are included:

```
seis4.su      output by acq4.sh (Section 6.3)
   27 Mbytes


Nshots.su     Nankai: 2-D line of shot gathers (Section 10.1)
  424 Mbytes     (from Prof. Greg Moore, U. of Hawaii)


Nstack.su     Nankai: stack 2-D line (Section 10.1)
   27 Mbytes     (from Prof. Greg Moore, U. of Hawaii)


Tshot.su      Taiwan: 2-D line of shot gathers (Section 12.1)
  411 Mbytes     (from Prof. Greg Moore, U. of Hawaii)
```

The following programs are included:

```
sukeycount.c
sudumptrace.c
tvnmoqc.c
```

The following makefile is included:

```
makefile
```

The following figures are included in the "figures" directory:

```
fig-07.5.tif          fig-10.8.top.tif       fig-13.6.top.tif
fig-07.6.tif          fig-10.8.bottom.tif    fig-13.6.bottom.tif
fig-07.8.tif          fig-10.9.top.tif       fig-13.7.top.tif
fig-07.9.tif          fig-10.9.bottom.tif    fig-13.7.bottom.tif
fig-07.10.tif         fig-11.1.left.tif      fig-14.1.tif
fig-07.11.tif         fig-11.1.right.tif     fig-14.2.top.tif
fig-07.12.right.tif   fig-11.5.tif           fig-14.2.bottom.tif
fig-08.1.tif          fig-13.1.top.tif       fig-14.3.tif
fig-08.2.tif          fig-13.1.bottom.tif    fig-14.4.top.tif
fig-08.5.tif          fig-13.2.top.tif       fig-14.4.bottom.tif
fig-08.6.tif          fig-13.2.bottom.tif    fig-14.6.tif
fig-09.5.tif          fig-13.3.top.tif       fig-14.7.tif
fig-10.5.top.tif      fig-13.3.bottom.tif    fig-14.8.tif
fig-10.5.bottom.tif   fig-13.5.top.tif       fig-15.2.left.tif
fig-10.7.tif          fig-13.5.bottom.tif    fig-15.2.right.tif
                                             fig-15.3.tif
```

The "figures" directory has all the color figures and a few others, all in uncompressed TIFF format.

E.2 Alphabetical List – Unix (excluding figures)

Ngazmig.sh	acq4shot.sh	makefile	seis4.su
Niviewcvp.scr	fxdisp.scr	migStolt.sh	shotrwd1.sh
Niviewcvp.sh	fxdisp.sh	migcvp.sh	shotrwd2.sh
Nmigcvp.sh	idecon.scr	migmovie.sh	shotrwd3.sh
Nmigmovie.sh	idecon.sh	model1.sh	shotrwd4.sh
Nnmo.sh	ifk.scr	model2.sh	showcmp.sh
Nshots.su	ifk.sh	model3.sh	showshot.sh
Nsort2cmp.sh	igain.scr	model4.sh	showshotB.sh
Nstack.su	igain.sh	myplot.sh	sort2cmp.sh
README	iva.scr	nmo4.sh	sudumptrace.c
Tmigcvp.sh	iva.sh	nmo4a.sh	sukeycount.c
Tmigmovie.sh	iview.scr	oz14prep.sh	tf.sh
Tnmo.sh	iview.sh	oz14velan.sh	tvQC.sh
Tshot.su	iviewcvp.scr	psmerge1a.sh	tvnmoqc.c
acq1.sh	iviewcvp.sh	psmerge1b.sh	velanQC.scr
acq2.sh	killer.sh	psmerge2a.sh	velanQC.sh
acq3.sh	killer.txt	psmerge2b.sh	
acq4.sh	killer2.sh	psmerge3.sh	

E.3 Alphabetical List – PC (excluding figures)

acq1.sh	iviewcvp.sh	nmo4.sh	shotrwd4.sh
acq2.sh	killer.sh	nmo4a.sh	showcmp.sh
acq3.sh	killer.txt	Nnmo.sh	showshot.sh
acq4.sh	killer2.sh	Nshots.su	showshotB.sh
acq4shot.sh	makefile	Nsort2cmp.sh	sort2cmp.sh
fxdisp.scr	migcvp.sh	Nstack.su	sudumptrace.c
fxdisp.sh	migmovie.sh	oz14prep.sh	sukeycount.c
idecon.scr	migStolt.sh	oz14velan.sh	tf.sh
idecon.sh	model1.sh	psmerge1a.sh	Tmigcvp.sh
ifk.scr	model2.sh	psmerge1b.sh	Tmigmovie.sh
ifk.sh	model3.sh	psmerge2a.sh	Tnmo.sh
igain.scr	model4.sh	psmerge2b.sh	Tshot.su
igain.sh	myplot.sh	psmerge3.sh	tvnmoqc.c
iva.scr	Ngazmig.sh	README	tvQC.sh
iva.sh	Niviewcvp.scr	seis4.su	velanQC.scr
iview.scr	Niviewcvp.sh	shotrwd1.sh	velanQC.sh
iview.sh	Nmigcvp.sh	shotrwd2.sh	
iviewcvp.scr	Nmigmovie.sh	shotrwd3.sh	